Business Innovation and ICT Strategies

Sriram Birudavolu • Biswajit Nag

Business Innovation and ICT Strategies

palgrave
macmillan

Sriram Birudavolu
Sr. Vice President - Information
Sciences
T-Hub Foundation
Hyderabad, India

Biswajit Nag
Indian Institute of Foreign Trade
New Delhi, India

ISBN 978-981-13-1674-6   ISBN 978-981-13-1675-3   (eBook)
https://doi.org/10.1007/978-981-13-1675-3

Library of Congress Control Number: 2018953332

Cover credit © kynny/iStock / Getty Images Plus

This Palgrave Macmillan imprint is published by the registered company Springer Nature
Singapore Pte Ltd.
The registered company address is: 152 Beach Road, #21-01/04 Gateway East, Singapore
189721, Singapore

PREFACE

Yat Bhavo Tat Bhavati (You Become What You Believe In)
Hence your reality is shaped by your beliefs and understanding.
                                    —*Sanskrit Saying*

Consider a typical current day scenario.

A mid-sized information technology services company has dozens of small- and mid-sized enterprises as clients, spanning many sectors—finance, health, logistics, academia, government departments, non-profit organizations, media, and so on. The IT company finds itself in the eye of the proverbial storm of the Information Age.

Each client has unique and changing requirements, owing to new exciting business models launched every month which solve interesting problems in the industry. Being deep into the Information Age, *everything* from entertainment and fun to customer support, consulting, and business seem to lean 100% on information and communications technology. Data pours in 24/7 for every source, from websites, mobile applications, desktop applications, bank ATMs, and IoT (Internet of Things) devices. Putting it mildly, the IT company and its clients are finding it rather dicey to keep pace with the changes in technology and business.

The design and software changes are taxing them to the hilt. Data security is a constant nightmare. There are threats, round the clock, of hacking, malware and exploits, and a few actual incidents as well. The underlying technologies in the server software and hardware are getting more complex by the day. And the demands for high performance, 24/7 availability, and instant access to every kind of data seem to be rising incessantly.

Technical expertise seems to be in *constant shortfall*. Budgets are very tight. There just aren't enough technical people to send to clients' places to understand and resolve the network (firewalls, core and secondary switches, Wi-Fi networks), hardware, and software problems. Regulatory compliance demands that each client meet architectural standards, maintain records, and meet data and transactions security and privacy standards for its end-customers.

How can an IT company not only crack these problems but also ride the changes and profit from them?

The IT company strategically attacks the problems with the following solutions:

- Get the clients to *move their data to the cloud*, that is, the IT company's data centers, thereby managing *in one place*, security, scalability, redundancy, and 24/7 availability.
- *Remotely manage* the clients' websites, portals, and mobile apps from an NOC (network operations center) and the security through an SOC (security operations center).
- With remote access to the clients' infrastructure (networks and servers), manage them *remotely over a secure VPN connection*, thus cutting 99% of the travel costs.
- The IT company's data centers reside in several cities, and they're all interconnected. So, the experts from any of the centers *can connect to any of the clients* all over the country and resolve problems. The *problem of shortage of experts is solved*, and their utilization improves.

With these in place, the next innovative steps through which the IT company can offer superior services are:

- *Predicting IT problems* intelligently from the trends and typical patterns, automating the model building, and taking preventive steps, and using data to validate and improve the models.
- *Helping clients improve their services* with their end-customers, with big data analytics, machine learning, and AI. For example, if the transaction traffic peaks in certain months, then add server capacity or catch fraud when the machine learning catches outliers.
- *Mining* the web-traffic data or mobile download patterns, AI tools can detect which type of visitors visit a client's website, how often, and from where (as IP addresses can reveal geography). The company

can suggest to the clients *to serve up different content* based on these variables, thereby improving the accuracy of targeting and reach.

- Build *a repository of patterns* for each area, from cyber-security to every client sector, develop a center of expertise, and recommend best practices and policy to the industry and regulatory bodies. Develop a high-end consulting practice and move up the value chain.
- Deploy state-of-art IT infra products to manage the bulk of the operations *autonomously*, including monitoring, optimizing, troubleshooting, auto-correction, and self-healing but also fault detection and resolving issues instantaneously.

Note that each of these innovative steps doesn't just solve problems, but they increase value and brand name, decrease costs and manpower, and lead to future growth for the IT company and its clients.

What is innovation? Let's start with a simple notion that you often know innovation when you see one. You can perceive innovation in uniqueness, whether manifest or subtle, in an element or in the blend, in the display or in the force, in the arrangement or in the essence, in the action or in the effect, in the grandeur or in the simplicity, in brilliance or in repose, in puissance or in stillness, in isolation or in association.

The myriad examples of outstanding innovation at work show that it is indeed possible, and this book will share with you the secrets of using ICT to create innovations. Yet, they are uncommon, and failures are many. There are great obstacles. The general disquiet that pervades the milieu stems from the confusing mix of elements, vigorously at play. The effects of ICT are all-pervading, and while the boons are many and mostly evident, jeopardy lurks at every corner. The triumphant organizations have crossed many difficult passes and have undertaken incalculable risks to achieve remarkable levels of success. It need not be so galling.

Opportunities abound everywhere, and creativity knows no bounds. ICT can and is being applied in eliminating labor, in communication ("geography is history"), in entertainment and the arts, in imbuing systems with intelligence and diligence, in multiplying power and efficiency, and in fine-grained analysis and synthesis. Why does it seem so difficult to realize value, then? For each success story, there seem several that have failed.

ICT has become astoundingly powerful and staggeringly complex at the same time. Enterprise investments in ICT are a leap of faith, in terms of the known and unknown risks undertaken. Both enterprise providers and enterprise consumers of ICT are caught in a tsunami of change that has swept away many businesses, both old and new. It is essential now to future-proof an organization's vision, strategy, and implementation for using ICT as an enabler, catalyst, and differentiator. The symbiotic process between the development of IT systems and the requirement of a modern-day organization has motivated us to think through to unfold this relationship and hence writing this book.

This book is devoted to the real process of unleashing the power of ICT and harnessing it for all future. It is intended to be a practical guide that is used on a daily basis. The book will cover all important aspects of ICT in a nutshell and show the path to the future through innovation.

What is the key to leveraging ICT for success in business for individuals, leaders, and organizations? Many answers come to mind, particularly ICT's amplified ability to serve customers by providing outstanding products and services and creating high value and doing it in ways that are effective, efficient, profitable, competitive, and valued.

How does one establish a business focus with ICT in mind? This is a key precursor to establishing an ICT strategy and is covered well in this book.

This book presents a lean and coherent framework distilled from years of research and experience in the ICT industry and business strategy development which captures the essence of some of the best practices used in the business, founded on a contemporary theoretical basis. What unites the different elements of the framework is their proven power and potency. Not only are they insightful and useful, but they have worked, often brilliantly and despite great adversity. The framework is built around these ideas and is based on Sriram's PhD thesis and key principles laid down in other industry-proven standards. The framework is enriched with Biswajit's long experience in research in the areas of economics of firms and business strategy development, especially his expertise in analyzing firms' behavioral change in a dynamic world. Open innovation, the quadruple helix model of innovation, enterprise architecture, ICT industry standards, regulatory environment, and the relevant firm theories are the cornerstones of the book. The process that runs the framework is driven by the themes of focus, integration, competition, regulation, change, and innovation.

The quadruple helix model involves collaboration with four different players—partner firms, research institutes, government agencies, and the users. Open innovation is about using external ideas that originated outside the boundaries of the firm as well as letting other organizations use some of the firm's ideas. External and internal economies of scale drive the firm in a competitive dynamic business environment. For the purposes of this book, the related concepts of open innovation, co-creation, network innovation, and collaboration have been clubbed into the term open innovation.

The first chapter, the *Introduction: Unfolding ICT*, lays the groundwork for the book. It also dispels any illusions that ICT is just another sector and the recent advances, a temporary development. ICT is the *zeitgeist*. The increasing bond between ICT and innovation has practically led to their fusion in many forms.

The second chapter, *The ICT Tsunami and Your Future*, while being cautionary, brings out the essence of the sweeping waves of changes precipitated by ICT and the innovations in its wake, some of them transforming ICT itself. The third chapter, *Every Organization Is a Complex ICT System*, is a continuation of the second and shows how every organization is ICT driven and why we need an "ICT lens" for looking at the modern and future landscapes of industry and society.

The fourth chapter, *The Enterprise Technology Landscape*, delves a little deeper into ICT itself, to get a clear understanding of the components and forces, and of how organizations and people are using it. Taming the wild forces of ICT needs strong governance, and this is also discussed.

It is easy to lose oneself in the immense complexity and the ever-shifting mess. The fifth chapter, *Finding Business Focus with ICT*, emphasizes the value of Focus for survival and growth, in this unbelievably turbulent ecosystem. It also shows a clear path to achieving and sustaining focus.

The sixth chapter, *The Innovation Strategy*, prescribes the essential ingredients required for formulating an effective ICT-based innovation strategy. This is based on vast research, industry experience, and evaluation of business trends.

Without the right people and culture in an organization, strategy is practically useless. The seventh chapter, *Social Capital: The Innovation Culture*, deals with the very important subject of developing the right culture in an organization to foster and turbocharge innovation.

Strategies and culture differ from region to region. The eighth chapter, *Regional Factors Influencing Innovation*, describes how an organization should factor in regional issues for developing a strategy for a region.

The ninth and tenth chapters focus on India. The ninth chapter talks about the current state of the ICT sector in India, a bit of regulatory environment, and some of the emerging issues India is facing internationally. The tenth one describes the evolving regulatory scene.

The eleventh chapter comprehensively sums up the ways in which to *Win the Competition with Innovation*.

The book drives home the concepts through apt mini-stories and anecdotes from the industry and attempts to develop an action plan on how to apply the relevant part of the framework. The framework and process emphasize Peter Drucker's point that great ideas and decisions are a blend of intuition and rigorous analysis.

The genesis of the book goes back to the days when both the authors had animated discussions regarding the ultimate use of a doctoral thesis for the people at large who wanted to know the dynamics of ICT and how it was useful for strategizing the business. The intense discussion on the brewing idea finally moved us to take the plunge in writing this book. The head of ICRIER, Rajat Kathuria, prompted Sriram also to convert the content of the PhD thesis into a book. Finally, it has become a joint product bringing the wisdom of two authors: experience of the ICT industry and research on business dynamics and strategy. The discussions with various experts and colleagues in India and abroad were extremely useful. It is difficult to thank all of them for their valuable insights on the topic. Nonetheless, the rich contribution is acknowledged with profound thanks. We would like to especially thank the faculty at the Indian Institute of Foreign Trade. Their support and inspiration have pushed us towards the early completion of this book. Professor O. P. Wali has been of tremendous help in being both a source and a sounding board for several ideas. Sriram would like to thank the management and colleagues (too many to name here) at Oracle Corporation and its customers and partners, for their valuable expertise and help during the time he was working with Oracle (2008–2017) in the Global Communications Business Unit. He gained much experience and many insights into the ICT and innovation during this period. Sriram is grateful to the colleagues and the management team at T-Hub foundation and the Director General of RICH, Ajit Rangnekar, for the valuable knowledge that they shared about startups, accelerators, startup ecosystems, and research forums and for the permission to use materials from T-Hub and RICH in this book. The market research analysts at T-Hub, Sandhya Kanukollu and Abhineeta Raghunath, have contributed significantly to the chapter on T-Hub, and Dr. Maneesh Kunte provided material on RICH. Sriram is also indebted to his family, particularly his wife Sivani, for putting up with his research and book writing for hours and days on end. Biswajit is also appreciative of his executive students who brought up lively discussions in the classroom on ICT and international business. Many of them are with startups or MNCs and trying to grapple with the speed of technology and its impact on business. The inputs helped us put the issues together, frame the chapters from the users' point of view, and identify the "Key Take Away" from each chapter. Biswajit is also thankful to his wife Bansari for her intelligent comments on some chapters and her ideas on the overall theme. We are also especially thankful to reviewers and our publisher for encouraging us to take this book project forward. We are indebted to many scholarly works and referred them wherever possible. As we submit our manuscript, we remain solely responsible for errors and omissions.

Hyderabad, India   Sriram Birudavolu
1st July 2018
New Delhi, India   Biswajit Nag

Praise for *Business Innovation and ICT Strategies*

"If you don't have a good innovation process, you have an idea graveyard. *Business Innovation and ICT Strategies* is a practitioner's handbook that reveals the secret behind the new idea economy."
—Vish Nandlall, *Head of Emerging Technology and Ecosystem Development at Dell EMC, Ottawa, Canada*

"Excellent congruence of ideas and practical methodology to harness the power of ICT to compete in a hyperconnected world. Highly recommend this book by Dr. Sriram Birudavolu and Dr. Biswajit Nag for organizations that struggle with internal lack of grasp in how the ecosystem is shaping up."
—Pavan Malladi, *Head of Governance, Strategy and Execution Excellence, Philips, Netherlands*

"Happy to announce that Dr. Sriram Birudavolu has just released his new book *Business Innovation and ICT Strategies* showing that the key to survival and growth for organizations is focus, innovation and creating value-webs at every level—from service delivery to building new business models."
—Tony Poulos, *Managing Editor, Disruptive Asia*

"I am confident this book will be incredibly useful not only for the MNCs but also for organizations in the informal sector who seek to thoroughly modernize and revamp their business, such as managing supply chain, improving quality and re-vitalizing human resources."
—Pawan Agrawal, *CEO of Mumbai Dabbawalas*

"Just finished a book titled *Business Innovation and ICT Strategies* by Sriram Birudavolu and Biswajit Nag. The authors used very simple, easy to understand stories and anecdotes to explain complex issues like Innovation, Enterprise Architecture from the prism of Business Value. A must read for professionals who wish to innovate in ICT while understanding the nuances of theories of economics. Happy reading!"
—Bharat Anand, *Chief of Technology, Ministry of Home Affairs, Government of India*

"The dramatic advances in technology have created huge and untapped opportunities to improve almost every facet of life and society. The barrier to capturing these opportunities is no longer the cost or availability of technology, but people and organizations. *Business Innovation and ICT Strategies* provides a practical set of ideas and frameworks for both enhancing innovation to identify value from technology and stimulating adoption by organizations. The book also recognizes the reality that adoption is increasing a multi-party ecosystem across companies, partners, and others."

—Eric Simonson, *Managing Partner of Research, Everest Group*

"*Business Innovation and ICT Strategies* is a very timely and relevant book which will help enterprise leaders as they embark upon the journey of Digital Transformation. The book provides very practical and easy to implement ideas and strategies to use Information and Communication Technologies (ICT) to digitize all aspects of the working and operations of an organization. The suggested frameworks are based on years of solid research by the author and backed by theoretical concepts. It demonstrates how to use Open Innovation and Quadruple Helix model to drive and implement ICT for deriving desired Business Outcomes."

—Vinod Sood, *Managing Director, Hughes Systique Corporation*

"A book which completely encapsulates the expertise and experience which the authors possess—an absolute amalgamation of industry and academia … a must read for all those who desire a comprehensive framework and a strategy for organizations and individuals related to ICT and Innovation."

—Yajulu Medury, *Director, Mahindra Ecole Centrale*

'*Business Innovation and ICT Strategies* by Sriram Birudavolu and Biswajit Nag is a rare piece of distilled practical ICT knowledge embedded in the open innovation framework that blends well with Agile principles. Agile breaks down organizational silos, Open Innovation (OI) takes this further by breaking down cross organization silos itself and promoting collaboration between organisation and customer and not the least also include the academia sector. The book however does not stop with the easier, less painful 'inbound' innovation stream, who would not like new ideas flowing into his organisation, but makes a real effort addressing the need to have a consistent outbound innovation stream in order to address the need to fertilize the ground for innovation. The authors exhibit a clear vision that could become one of the cornerstone pieces in the disruptive world especially in combination with distributed ledger (blockchain) incentive models that might be the missing piece to address the problem of IP/Knowledge ownership."

—Roland Pfeiffer, *Former Head of New Technology and Service Flows,*
Celcom Axiata Berhard

"With the digital transformation in full swing, Birudavolu and Nag's book clearly describes why and how to harness ICT to increase competitiveness and innovativeness. His findings are seminal and practical, and a must read for leaders on their journey to create and capture value in innovation ecosystems."

—Dr. Daniel Fasnacht, *Founder/CEO, EcosystemPartners Ltd., author of the book* Open Innovation Ecosystems, *Former CEO/Managing Director of Swiss Banks and Financial Institutes—Julius Baer, Kaiser Partner, Tom Capital, VP in Credit Suisse*

"ICT emphasizes the importance of the media in the use of information technology, that should be considered through a social, behavioral and environmental approaches. Being in the forefront of meeting technology disruption, on a daily basis, in many countries worldwide, I find Dr. Sriram Birudavolu's book to be of value and importance to for doers, customers and decision makers, to get up close and personal with ICT impact and its effects through open innovation and other practical means. Enjoy."

—Tal Catran, *Accelerators Guru & International Keynote Speaker, Start-Up Ecosystem Builder, TEDx Speaker, Israel*

"Business Innovation and ICT Strategies is an enlightening book written by Dr. Sriram Birudavolu & Biswajit Nag. The book focusses on the real process of unleashing the power of ICT and harnessing it for future and the best part is that it can be referred daily to further your strategic and tactical actions around business innovation using ICT's disruptive power. The frameworks in the book are backed by rich research which is visible when you are able to create an action plan after every chapter using those frameworks. It is definitely a must read for business leaders and ICT professionals."

—Puneesh Lamba, *Group CIO, CK Birla Group, India*

"Innovation and Digital Transformation propelled by new developments in ICT will determine the survivors and winners in the coming decade. Dr. Sriram Birudavolu's experience driving innovation ground-up and his methodology for companies to adopt this is an essential read for the future industry leaders"

—Vijay Pullur, *Co-Founder, Pramati Group, and Serial Entrepreneur*

CONTENTS

About the Authors

Sriram Birudavolu is Senior Vice President and Head of Information Sciences at T-Hub Foundation, India's largest hub for technology start-ups. He also runs the market research program at T-Hub. With over 24 years of experience in the ICT/telecom industry, he is a C-Level advisor and has worked on four continents in over a dozen countries, with reputed organizations, such as Oracle, Hewlett-Packard, Verisign, and Siemens. His areas of interest are business and technology innovation, ICT, IT, and telecom and enterprise architecture. He holds a PhD in international business from the Indian Institute of Foreign Trade (specializing in open innovation in ICT), a master's in computer science, and a bachelor's degree in electrical engineering. He is a TOGAF certified enterprise architect and a certified PMP (project management professional). He has also published and presented papers in international journals and conferences and has delivered guest lectures in academic institutes and companies. Currently, he is an academic advisor to Mahindra Ecole Centrale Engineering College, Hyderabad, Indian Institute of Foreign Trade, Delhi and University College of Engineering, Osmania University, Hyderabad.

Biswajit Nag is Professor of Economics at the Indian Institute of Foreign Trade (IIFT), New Delhi. Involved in empirical economic research for close to two decades, he also has teaching experience in India and abroad. Earlier, he served the Poverty and Development Division of UN-ESCAP, Bangkok. Dr. Nag has completed a number of projects for the Government of India and international agencies such as the UN, World Bank, Asian Development Bank, WTO, DFID, and the EU. His current research interests are trade and technology, international production network, global value chain, trade in services (including IT-enabled services), and so on. Dr. Nag is one of the advisors on global value chains at the Asia-Pacific Research and Training Network (ARTNet), UN-ESCAP. He is an active member of GERPISA, a Europe-based international automobile research network. Dr. Nag teaches international economics, global business environment, and econometrics at IIFT and provides his consultancy and advices to firms and government agencies on international market access and strategic issues.

LIST OF FIGURES

LIST OF TABLES

## CHAPTER 1

Introduction: Unfolding ICT

Sahase Srih Vasati (Luck dwells in the best of one's endeavor).
—Sanskrit Saying

1.1 The ICT Phenomenon

Information and communications technology (ICT) is permeating and transforming every sphere of life, including commerce, education, health-care, governance, agriculture, manufacturing, banking, defense, media and entertainment, logistics, travel, tourism, and so on. Organizations such as the World Bank and the Asian Development Bank have noted that telecommunications is now essential to a country's economic development and competitiveness. As per a World Bank Report, a 10% increase in the speed of high-speed Internet connection has been found to increase economic growth by 1.3% (Minges 2016).

Attractive as all this may seem, the forces unleashed by ICT are in fact so powerfully disruptive and rapid that organizations not only struggle to harness the potential of ICT but are also getting run over by the change, complexity, and competition. Complex regulatory structures in various countries and the trade dimension (such as service and manufacturing interdependence through equipment and software interface) have added their own obstacles in the path of harnessing ICT's potential benefits.

Where does one start? Not surprisingly, with the psyche.

© The Author(s) 2019
S. Birudavolu, B. Nag, *Business Innovation and ICT Strategies,*
https://doi.org/10.1007/978-981-13-1675-3_1

A shift in mindset is crucial not only to deal with these disruptions but also to exploit them and thrive towards greater successes. The shift in mindset is all about moving away from security and stability towards resilience. In contrast to the former approach which is more risk-averse, the latter indicates an acceptance of the inevitable reality that there will be shocks and disruptions and prepares for them, and hence actively engages in experimentation and risk-taking. It is a far more opportunistic and entrepreneurial outlook.

The opportunities and disruptions are equally immense. It is imperative to take a closer look at both.

The global mobile devices and connections worldwide today exceed the world population (7.25 billion). The mobile phone penetration is more than 90% and several people own multiple subscriptions.[1] In comparison, the PC penetration stands much lower at 20%. The total cost of computing processor power, bandwidth, and storage continues to drop (roughly halving each year), and newer, more efficient standards of network technology are continually being implemented.

All these have effected radical changes in the ICT landscape. Read carefully through the sweeping list. Each of these represents opportunity as well as disruption, depending on your point of view.

1. *In Services and Business Models*: immersive user experience including AR/VR/XR, digital services, digital marketing, analytics, big data, automation, enterprise architecture, microservices, open innovation, crowdsourcing, social media, e-commerce, and so on
2. *In Mobile Networks*: mobile broadband, 3G, 4G/LTE, 5G, end-user mobility (e.g. location-based services and mobile banking), enterprise mobility (e.g. field personnel enablement and logistics management), Telco mobility (e.g. smartphones and smart applications), and so on
3. *In Infrastructure and Data Centers*: cloud computing, virtualization, containers, and services, for example, SaaS, IaaS, PaaS, KaaS, which mean software/infrastructure/process/knowledge as a service, and so on
4. *In Core Networks*: IP networks, SIP Trunking, SDN (software-defined networking) and NFV (network function virtualization), convergence (VoIP, triple/quad play), and so on

[1] Ericsson Mobility Report, 2016, www.ericsson.com.

5. *Regulations*: de-regulation, Net Neutrality, spectrum auctions, laws related to competition and monopoly, security of data, and so on
6. *Technology*: platform, smart machines, 3D printing, IoT (Internet of Things), Artificial Intelligence, Machine Learning, AR/VR/XR/MR, and so on

Upon adopting a new mindset of resilience, the next step then is to look at things through the lens of ICT. Every organization is a complex information engine and hence an ICT organization. Having established this, it is crucial to understand technology and enterprise architecture well. The key industry trends and the technologies of interest must be understood by everyone who plays any significant role in the organization. Technology is really an ecosystem enabler, and it will be beneficial for everyone to grasp this fact.

1.2   WHAT IS ICT?

An OECD (the Organisation for Economic Co-operation and Development) paper (Inaba and Squicciarini 2017) defines ICT as any product or service that stores, retrieves, manipulates, processes, transmits, or receives information electronically in a digital form. It is the convergence of unified communications, telecommunication, computing, and broadcasting industries. The older Information Technology (IT) is a narrower term which usually excludes telecommunications (voice) technology while including data networks (although almost all networks today are digital), as a reference to the systems that support information processing. The term Information Technology is used in a narrower sense, typically excluding telecommunications (voice) technology while including data networks (although almost all networks today are digital), as a reference to the systems that support information processing.

Processing (computing power), network, and storage capacities are rising exponentially, and knowledge is becoming accessible to more people than ever before in human history. Network effects trigger an ever-increasing number of people (and hence devices) to use and create information.

We're now into the *Fourth* Industrial Revolution (Schwab 2017) which will bring a thorough and exponential transformation in the industry. To clarify:

- The First Industrial Revolution was powered by steam-powered machines to drive production.
- The Second Industrial Revolution was powered by electrical machines in the industry.
- The Third Industrial Revolution brought in digital transformation in the industry.
- The present one, Fourth Industrial Revolution, is about integration and fusion of the physical, digital, and biological technologies.

Each of these industrial transformations was built on top of the previous one. Thus, the systems of the Fourth Industrial Revolution leverage the third, digital revolution, that is, global, digital communications, high-speed and low-cost processing, and high-density data storage, and a vast, digitally connected global population of people and devices.

Of the ten key emerging technologies that are among the most promising and potentially most disruptive and that carry significant risks outlined in the OECD science, technology, and innovation outlook (Inaba and Squicciarini 2017), four directly come under ICT (viz. Internet of Things, Artificial Intelligence, big data analytics, and blockchain), and the remaining rely heavily on ICT for their research/design/production/use (micro/nano satellites, additive manufacturing, neurotechnologies, advanced energy storage technologies, nano materials and synthetic biology). Similarly, PricewaterhouseCoopers has identified eight essential technologies for the future (PwC 2016).

Artificial Intelligence, IoT (Internet of Things), VR (Virtual Reality), AR (Augmented Reality), blockchain, robots, 3D printing, and drones.

All of these are ICT technologies.

In short, the disruptions and innovations in and driven by ICT are set to drastically change the way firms, industries, and economies work.

The future holds tremendously high potential for the development of individuals and organizations, due to the synergies among these technologies and also other upcoming technologies such as quantum computing, LiFi (Light Wi-Fi), nanotechnology, bio-engineering, and so on.

The present day's unified communications technologies such as SMAC (social media, mobility, analytics, and cloud computing) catalyzed by underlying technologies such as virtualization, container management, and microservices architecture are accelerating the velocity of business by the day, that is, speed as regards what firms can do, how they can do it, and how quickly they can deliver.

Fundamentally, ICT has magnifying power. ICT bestows on us the capability to gather and use intelligence at a rate and level unprecedented in history, as follows:

Collect big data from various sources, such as IoT-enabled devices, analyze this data as rapidly as needed by using analytics, mining, AI (Artificial Intelligence), and Machine Learning. The deep insights drawn by using these technologies to collect, analyze, and understand data give a huge opportunity to understand retail and enterprise customers, to fine-tune processes, cut inefficiencies, redundancies and wastage, enable e-governance, and conduct research and development. This entire process can happen in real-time too.

Gartner vice president Steve Prentice says that "At a minimum, the IT organization needs to be able to design the 'big picture' of all the new information and technology capabilities required to support digital business. ... Fueled by data, analytics and AI, algorithmic business will continue to grow and disrupt your business" (Gartner 2016).

1.3   ICT and Innovation

ICT spurs innovation at every stage of the lifecycle of a product or service. For research in the academia and industry, ICT tools facilitate both fine-grained accuracy in analysis and massive modeling and crunching capability to design, build, and validate complex theoretical models, by using big data, small data, or even imperfect data. It also enables geographically disperse teams (even crowds) to collaborate (Birudavolu 2015).

ICT helps in market research, ideation, and creation of new products and services. During the sales process, the customer requirements can be mapped and matched exactly to help craft solutions and proposals. With the richness of data (even if disperse/distributed) and the network of people and devices available to organizations, entirely fresh business models can be conceived. Existing business models can be made sharper and competitive through re-engineering.

In development and production, ICT enables seamless and agile processes, DevOps, smooth delivery, and operations of products and services. Exceptional quality, cost, customer experience, and other competitive advantages are obtained.

Talent development and learning/training are made vastly easier through guided learning with online courseware and simulations. Even systems are imbued with artificial intelligence so that they can learn in a supervised or unsupervised manner. This multiplies the power of automation and execution, so that the systems can function in a variety of different complex environments, not merely the kind of routine scenarios that conventional automation can handle.

In both direct and indirect ways, digital technology drives innovation by incentivizing all players to innovate. Without even having to invent new technologies, the markets become competitive on innovation. As per the World Bank's Global Information Technology Report, 2016, technology can impact the market in the following ways (Ballar et al. 2016):

1. *Increasing Market Size*: Technology integrates markets through superior, frictionless, instant, and fine-grained communication. This makes the markets competitive. For example, online platforms on which firms can reach a global customer base at a minimal cost.
2. *Reducing Entry Barriers*: Online and cloud services enable startups and SMEs (small- and medium-sized enterprises) to launch their businesses online at a very low cost. This reduces entry barriers and allows scaling, thus creating a level playing field. It is essential therefore to retain Net Neutrality through regulations.

   Mettler and Williams (2011) identify six such types of business platforms: crowd-financing, digital utilities, professional services marketplaces, micro-manufacturing, innovation marketplaces, and e-commerce platforms.

   Gartner (2016) identifies five major platforms of digital business as information systems, customer experience, data and analytics, IoT (Internet of Things), and ecosystems.
3. *Acquiring and Leveraging Knowledge of Consumer Preferences*: ICT enables matching opportunities to the products/solutions, by using big data to unearth customer needs and preferences. With customer centricity comes a greater customer experience and higher value delivered at a lower cost.

The importance of understanding, leveraging, and managing these disruptions is hence becoming ever significant for the success of each economy, industry, and organization.

Following Afuah (2003) innovation can be explained through a new product or services which comprises of either new technological knowledge or new market knowledge or both. This depends on competencies and assets an organization possesses. A company's ICT infrastructure, its skilled personnel, and tendency to develop strategies based on information available through ICT framework and so on build up its competencies and intangible assets which are essential for developing new product and services (Fig. 1.1).

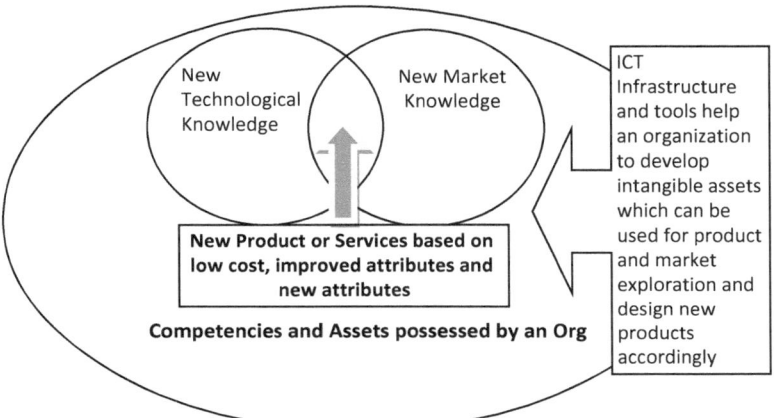

Fig. 1.1 ICT and innovation (Note: The above diagram is modified from the innovation structure described by Afuah 2003)

Innovation can be described in various ways with different structural differences. The following are some examples:

Incremental vs. Radical: The experience of new firms and old incumbent firms adopting incremental or radical innovation is mixed. Companies like GE an incumbent in medical diagnostic equipment industry did fairly well in adopting radical change in technology and moved successfully from x-rays to CAT scan to MRI. On the other hand, Intel struggled to move from one generation to next in the chip industry. Hence, we require more information to understand when incumbent embraces radical or incremental technologies.

Market vs. Technical Capabilities: The literature in such cases argues that even when a firm is not able to cope with technological capabilities, it can still remain at the top if it has more market knowledge and capabilities to reach customer. It may be able to remain ahead of any technologically superior firm if it possesses better market information. This provides a direction to understand why some firms with regular pace of innovation are still ahead of others who have niche or radical technologies. ICT tools are powerful to get more market information and consumer insights. This model is popularly known as Abernathy and Clark model.

Architectural Innovation vs. Component Knowledge: A successful innovation requires the knowledge of the components and the architecture of linking them efficiently. An innovation can impact either of the two or both. If the innovation brings newness in both component and architecture, it is *radical* in nature. If more focus is on the newness of components and bit of change in architecture, it is called *modular* innovation. It could with complete focus on *architectural* innovation where focus is on newness in architecture. Sometimes, companies take a up a *gradual/incremental* approach where both component and architectural knowledge is extended. Historically, it is observed that firms may have possessed good architectural knowledge but lack of component knowledge has inhibited them from success in innovation. ICT or digital technology provides a wonderful opportunity to firms to both working on architecture especially in developing design and components as per the need of the market. We can provide the example of Apple products in this context.

Innovation to Streamline Value Chain: Apart from product and process upgrading, ICT tools can be effectively used for functional and inter-chain upgrading which are an essential part of value chain streamlining. In case of functional upgrading, companies reduce the value chain, and there may be specialized suppliers who are ready to take charge of some parts of the value chain, and thereby do some agglomeration. For example, a product can be divided into different modules, and different companies manage those modules. In case of inter-chain upgrading, two or more different value chains are linked for new value chain. For example, value chain of basic electronic industries can be linked with toy industry to develop a new value chain for electronic toys. The role of ICTs in managing or streamlining value chain is immense. Walmart or Amazon's supply chain management could be good examples of how ICT can help in supply chain innovation.

The Takeaway Box: ICT and Innovation Are Keys to Your Future!

- Take ICT very seriously, regardless of the business you are in.
- Innovation is a huge driver for all businesses. You cannot continue to rely on past, static models and successes.
- In the constantly changing world, the paradigm of customer centricity is a given.
  Re-engineer and design all business processes around your customer needs.
- Turn to the different chapters in this book again and again to help clarify your thinking and address different parts of the puzzle, for your organization and your career.

1.4   THE ORIENTATION AND STRUCTURE OF THE BOOK

This book is committed to finding a real path through this complex web so that all organizations (not just Telcos) which are looking at leveraging/optimizing ICT for their business get a clear sense of direction. For this purpose, it is essential to first distil and capture the critical challenges on the path to leveraging ICT for success. Today's business innovation is dependent on tomorrow's vision of how ICT will help an organization grow. Hence, an organization needs to plan its strategic path for innovation today itself.

First, it is important to find the business focus in an organization (Ries 1996). Without focus, any strategy, ICT or otherwise, is doomed to failure. For example, many Telecom companies are confused as to whether they should function as a pure data pipe or become a media/content company, and this has led them to make unprofitable acquisitions. Another example is that of several enterprises, which are getting defocused due to competition and are taking half-baked decisions on balancing between how they sell through the new digital channels versus their traditional network. This has damaged their existing dealer network, with disastrous consequences for their business. And they have not managed to ramp up their online business also, due to a fragmented approach.

At the heart of the vast and complex ICT industry are a small set of core concerns such as *focus, integration, competition, regulations, change,* and *innovation.* These need to be addressed effectively through a comprehensive and integrated framework that is driven by a paradigm. It is essential to use a framework in order to avoid a piece-meal approach.

The leadership needs to build a strong strategy to realize its vision and powerfully bind all the key elements. Of special importance is the innovation component.

Open innovation is a paradigm that assumes that firms can and should use external ideas and internal ideas, and external and internal paths to market, as they look to advance their technology (Chesbrough 2006). The concepts of open innovation have been applied successfully in several industry sectors. The ICT domain is now ripe for employing these principles. For the sake of simplicity, in the term open innovation we also include co-creation (Ramaswamy and Gouillart 2010) and network innovation (Nambisan and Sawhney 2008).

From the perspective of management and technology, this book presents an integrated direction to a future with ICT and specific steps on the

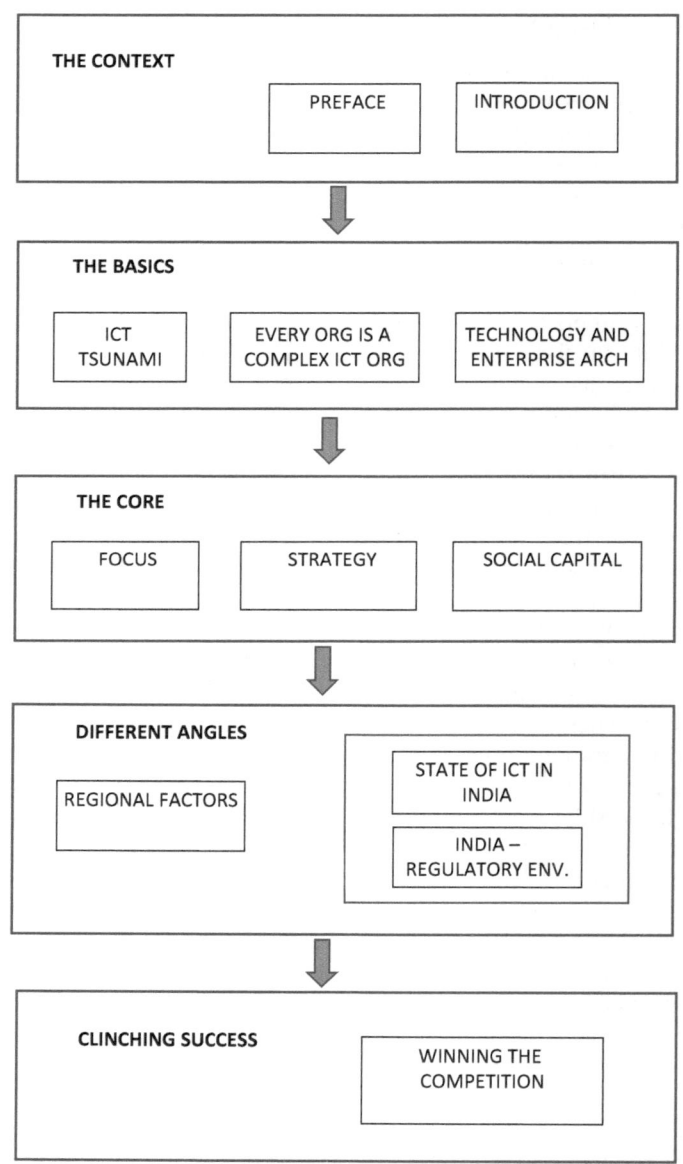

Fig. 1.2 Structure of the book

path. In this connection, the implementation of the Quadruple Helix Model, driven by open innovation, with four kinds of collaborators, comprising partners (that include other firms, such as competing Telcos), research institutes, government agencies, and the users, is important. Fundamental to this discussion would be the recognition that constructing the right strategy and building the necessary social capital are essential steps in such a process-oriented framework. How can organizations find focus using ICT and open innovation? This is the theme of this book.

An organization cannot realize its strategy without a solid organizational culture in place. This social capital is the real asset of an organization.

Strategy and social capital are both heavily influenced by regional factors. What works in one region may not work in another. On the other hand, a good understanding of regional factors can contribute towards making a product global and greatly successful. These issues have been covered in the book with number of examples.

India is one of the largest and certainly the most diverse economies in the world. A good understanding of Indian ICT can help every organization because there are many parallels, across the globe, in several different aspects. For example, the complex and evolving regulatory structure of Indian ICT offers many lessons for both regulators and organizations around the world.

How does one put it all together to craft a unified and successful game plan and win the competition? There are simple frameworks and processes that help in doing this, as described in the later chapter, "Winning the Competition".

Finally, two live examples of collaboration and innovation hubs are T-Hub and RICH. These are detailed in the last chapter.

The structure of the book is summarized in the following diagram (Fig. 1.2).

References

Afuah, A. (2003). *Innovation Management: Strategies, Implementation and Management* (2nd ed.). Oxford: Oxford University Press.

Ballar, S., Battista, A. D., Dutta, S., & Lanvin, B. (2016). Global Information Technology Report 2016, Chapter 1.1: Network Readiness Index. *World Economic Forum.* Retrieved from http://www3.weforum.org/docs/GITR2016/WEF_GITR_Chapter1.1_2016.pdf.

Birudavolu, S. (2015). Open Innovation in ICT: An Empirical Assessment of Global Telecommunications Services. Unpublished thesis submitted for PhD at Indian Institute of Foreign Trade, New Delhi.

Chesbrough, H. (2006). *Open Business Models: How to Thrive in the New Innovation Landscape.* Boston, MA: Harvard Business School Press.

Gartner Report. (2016, September 28). *Gartner Says New Digital Disruptors Demand a New Approach to IT.* Retrieved from https://www.gartner.com/newsroom/id/3458217.

Inaba, T., & Squicciarini, M. (2017). *ICT: A New Taxonomy Based on the International Patent Classification.* OECD Science, Technology and Industry Working Papers, 2017/01, OECD Publishing, Paris. https://doi.org/10.1787/ab16c396-en.

Mettler, A., & Williams, A. D. (2011). *The Rise of the Micro-Multinational: How Freelancers and Technology-Savvy Startups are Driving Growth, Jobs and Innovation.* Brussels: The Lisbon Council.

Minges, M. (2016). *Exploring the Relationship between Broadband and Economic Growth.* World Development Report 2016. World Bank. Retrieved from http://pubdocs.worldbank.org/en/391452529895999/WDR16-BP-Exploring-the-Relationship-between-Broadband-and-Economic-Growth-Minges.pdf.

Nambisan, S., & Sawhney, M. (2008). *The Global Brain: Your Roadmap for Innovating Faster and Smarter in a Networked World.* Upper Saddle River, NJ Wharton School Publishing.

PwC, Price waterhouse Coopers. (2016). *The Essential Eight Technologies: How to Prepare for Their Impact.* Retrieved from www.pwc.ru/en/8technologies.

Ramaswamy, V., & Gouillart, F. (2010). *The Power of Co-creation.* New York, NY: Simon and Schuster.

Ries, Al. (1996). Focus: The Future of Your Company Depends on It. *New York.*

Schwab, Klaus. (2017). The Fourth Industrial Revolution. *Crown Business,* ISBN-10: 1524758868, ISBN-13: 978–1524758868.

## CHAPTER 2

The ICT Tsunami and Your Future

There is a tide in the affairs of men.
Which, taken at the flood, leads on to fortune.
Omitted, all the voyage of their life
Is bound in shallows and in miseries.
On such a full sea are we now afloat....
—*Shakespeare (Act 4, Scene 3, Julius Caesar)*

2.1 Introduction

The forces unleashed by ICT and innovation have kicked off an unending series of enormous hurricanes in corporate and non-corporate landscapes across the globe. As the quote above indicates—while the tidal waves of these tsunamis present unprecedented challenges, there are also gigantic opportunities waiting to be seized. Ignoring these challenges is a prescription for heavy defeat for both organizations and individuals. This is the exact situation we're in now.

This chapter outlines the nature of the destructive forces, and how they can result in a complete disaster for organizations.

As per Innosight's 2018 study on corporate longevity:

- Over the last 50 years, the average lifespan of the Standards & Poor's 500 list of companies has shrunk from 60 to 18 years (Innosight 2018).
- Half of the existing S&P 500 companies are expected to be replaced within the next ten years.

© The Author(s) 2019
S. Birudavolu, B. Nag, *Business Innovation and ICT Strategies*,
https://doi.org/10.1007/978-981-13-1675-3_2

As per a study done by Capgemini Consulting on digital disruption (Capgemini 2015):

- Fifty-two percent of the Fortune 500 have been merged and acquired or have gone bankrupt since 2000.

As per (Perry 2017), the public policy think-tank body AEI (American Enterprise Institute):

- Less than 12% of the Fortune 500 firms remain today when comparing 1955 vs. 2016.
- The speed of disruption is accelerating: In the period 1955–1994, about 8.5 new firms per year were added into the list, as compared to 14.2 new firms added per year in the period 1995–2016, which is almost double the rate.
- Forecasting these rates, the entire Fortune 500 may be replaced in about 25 years, that is, by 2045.

If this is the fate of giant corporations with war chests of billions of dollars at their disposal, then what future awaits the others? As we know, the larger ships sink more slowly, whereas the smaller ones may be wiped out practically overnight.

As per former Citigroup CEO Vikram Pandit, around 30% of banking jobs may disappear in the next five years, due to developments in technology. He said, "Everything that happens with artificial intelligence, robotics and natural language, all of that is going to make processes easier" (Livemint, September 2017).

The new waves of technology, radical business models, competition, and regulations which hit the industry at an unrelenting and ever-increasing pace are ample illustration of the fact that ICT has become all pervasive in transforming all the industries. Given this context, it becomes imperative for organizations to understand the main elements of such change(s) especially when they have decided to embrace the ICT models. Concomitantly, this will also enable them to effectively map out their future based on how relevant each of the specific elements are to their location/profile(s). This chapter will therefore, explore the manner/ modalities through which ICT is acting as a major force of transformation within the industries.

Read through the following facts to understand the full impact of the transformation that ICT brings:

1. The US regulatory FDA (Food and Drug Administration) has approved Viz.ai to use its AI algorithms in the market to help doctors and hospitals quickly flag stroke cases for patients admitted to emergency rooms, thereby enabling rapid treatment (Simonite 2018). Viz.ai automatically analyzes the CT scans for stroke patterns. Time is precious as brain tissue dies every second after a stroke.
2. Microsoft's Speech Recognition achieves the lowest error rate of 5.1% in August 2017 that puts its accuracy on par with professional human transcribers, who have an error rate of 5.9% (Microsoft 2017).
3. Just 14 employees make 500,000 tons of steel a year in Voestalpine AG's plant in Donawitz, Austria (Biesheuvel 2017). The unit is so highly automated.
4. GE has 3D printed a working jet engine (Keller 2016). Aircraft jet engines are considered one of the most critical and highly complex engineering products. In October 2017, GE Aviation completed successful tests of FATE engine and T901 engine prototype, both featuring 3D printed components (Saunders 2017).
5. Rolls-Royce Trent XWB-97 completes its first test flight in November 2015; it is the world's largest 3D printed aerospace structure, and it will power the long-range A350-1000 aircraft (Rolls-Royce 2015).
6. The most valuable retail company in the world sells entirely online—Amazon has a market cap of $440 billion. The second most valuable retail company in the world, Alibaba, with a market cap of $392 billion, is again an entirely e-commerce company, and additionally it does not own a single store or warehouse—Alibaba posted a revenue of US$23 billion in 2017 (Mourdoukoutas 2017) (Wikipedia Amazon 2018/Wikipedia Alibaba 2018). And both of them are highly automated.
7. Amazon USA needs only a minute of human labor to ship a package. The remaining work is done entirely by robots and automated systems (McFarland 2016); Amazon has 45,000 of them in 20 warehouses. Alibaba's logistics arm JD-X has a fully automated, dark warehouse in Shanghai that is 3 football fields big (16,000 square meters), in which robots work round the clock even in darkness (Xia 2016).

8. ORNL (Oak Ridge National Laboratory) 3D printed a house and a car that produce and share clean energy (Kira 2015).

9. BioCarbon Engineering, a UK-based startup, wants to restore earth's environment on a war footing and on an industrial scale. It uses drones to plant trees rapidly and reforest large expanses of land. Its drones scan the land for suitable soil and fire seedpods with enough force for them to break the surface. So far it has planted 25,000 trees (BioCarbon Engineering 2018).

10. The largest hospitality company in the world does not own a single room—Airbnb website has over 4 million lodging listings in 65,000 cities and 191 countries and made a revenue of $2.6 billion in 2017 (Airbnb 2018).

11. In May 2017, Google launched its AutoML project, which is AI (Artificial Intelligence) software that helps human AI engineers to build other AI software. By October 2017, Google announced that the AutoML software was building better AI software than the human engineers themselves, with regard to tasks ranging from simple (classifying images based on content) to complex such as marking location of multiple objects in an image (Fossbytes 2017).

12. The largest taxi company in the world does not own a single taxi—Uber operates in 633 cities worldwide, with a revenue of US$7.5 billion in 2017, a year in which it powered 4 billion rides (Forbes 2018).

13. Norway plans to launch the world's first autonomous and fully electric ship carrying 100 containers in 2018, which will cut operational costs by 90%. It will save 40,000 truck journeys per year. After testing, the ship will be completely crewless by 2020 (Paris 2017).

14. As of February 2018, there are at least 45 *mobile-only* banks in the world (Mobile-only Banks 2018). These banks operate without branches, and only via a smartphone (Mobile-only Bank 2018).

15. The largest shoe company in the world does not own a single factory—Nike drew revenues of $34 billion in 2017 (Nike 2018).

15. The largest television network in the world functions entirely online and does not own a single broadcast station—Netflix with 117 million subscribers globally in January 2018 drew revenues of $11.7 billion in 2017 (Netflix 2018).

17. The biggest education/training organizations in the world do not own a single classroom—Coursera and edX deliver MOOC (massive online open course). Coursera has 23 million subscribers and 1700+ courses; edX has 10 million subscribers and offers 1270 courses online (www.coursera.com, www.edx.com).

18. In 2011, Facebook has commenced operations in its massive Arctic data center, about the size of six football fields, located in the north of Sweden, in the Arctic circle (to save cooling costs). The level of automation is so high that there is one employee required for every *25,000* servers (Weiner 2016).

Every business and activity, whether for profit or non-profit, and regardless of the sector, stands to get disrupted. Business cycles are getting faster than ever before, as reflected in the vanishing of companies from the S&P 500 list.

The sweeping shifts in the industry caused by ICT and innovation are summarized in the following table (Table 2.1). In fact, the table should serve as a comprehensive checklist to measure the extent to which your organization and you yourself are in tune with the paradigm shifts.

Rate your organization on a scale of 1–10 on each of the rows, relative to the second and last columns. The lower end of the scale (1) lies in the second column and the high end (10) in the third column. If the total score, added up for all the rows, depicts a picture that largely represents the second column, that is, it falls short of the last column by a sizeable margin, then the organization is quite out of step with the times. An organization that scores low on "change quotient" is at very high risk and warrants immediate planning towards digital transformation. First and foremost, it's a question of survival. The changes may hit the organization in any form at any time and will demand urgent action. Second, these changes deserve to be fully exploited and to thrive, not merely survive.

2.2   A Connected World Poses Challenges: Contagion Effects

ICT has helped in getting the world remarkably connected as never before. The network effects are indeed non-linear and extraordinarily powerful. However it is also a double-edged sword; the disturbances in one part of the world have a contagion effect on the other parts practically immediately.

Table 2.1 The major shifts summarized

#	Shift in the paradigm of	Shift from	Shift to
1	Infrastructure	Hardware	Software (code)
2	Delivery mechanism	Products	Services (with contractual SLAs)
3	Granularity of delivery (touch points)	Basic services closely tied to the products delivered	Microservices flexible; mix-and-match as the situation demands
4	Engagement stage	Downstream	Upstream
5	Deployment duration	Months	Minutes
6	Nature of processes	Traditional, siloed	Agile across the organization from sales to DevOps
6	Performance speed	Hours/minutes	Milliseconds
7	Scale of transactions per second	Dozens	Millions
8	Availability	Working hours or not defined	24/7 with 99.999% uptime
9	Reach	Local	Global
10	Ecosystem boundaries	Closed	Open
11	Connectedness	Isolated	Networked
12	Location	Co-located	Distributed/cloud
13	Design	Isolated/fragmented	Integrated and flexible Internet and cloud based
14	Architecture	Monolithic	Components based, fine-grained, scalable, layered, and segmented
15	Interfaces between internal systems and with external systems	Rigid and limited	API-based contract-driven (application programming interface)
16	Security	Limited security	Highly secure at every level built for security from the ground up
17	Automation (development, operations, and all processes)	Very limited	Completely automated with very few exceptions

18	Intelligence	Extremely limited. Just some business reports	High degree of intelligence, with analytics, Artificial Intelligence, and Machine Learning in every department. Deep knowledge bases
19	Competitive	Limited awareness of trends in competition. Legacy outlook in competitive positioning and marketing strategy	High degree of awareness, and competitive positioning in the market, even from unconventional competitors. Always in tune with constantly changing trends. Dealing with hyper-competition
20	Product/service offerings, or vendor procurement	From thousands to millions of dollars upfront	In cents per hour, or even free (esp. for digital products), freemium
21	Innovation	Incremental and internal to the organization	Disruptive and open; collaborative across orgs and departments; quadruple helix
22	Technology used	Proprietary and uniform	Varied (open + proprietary + legacy + new). High use of open source
23	Usage model	Ownership or licensed	Subscription model, pay-as-you-go, pay as per usage, scalable
24	Compliance for regulatory, industry standards, contractual, operations, SLA, process, quality, architecture, and organizational compliance purposes	Extremely limited capability. Cannot show compliance in tune with rapidly changing regulations and standards	Ground-up compliant. Modular and flexible to build and show keen compliance at every level and at any time. SOC for security, risk and compliance, and ITIL, TOGAF, OWASP, and so on
25	Partnerships, JVs, and collaborations	Fixed set of few partners, long-term collaborations	Broad spectrum of collaborations of a large number and variety. These include long-term partnerships, mass collaborations, transient associations, crowdsourcing, open innovation, collaborations with end-users, competitors, vendors, customers, and so on
26	Network effects	Limited and linear	Powerful and non-linear play. Each piece of work has the potential to disrupt the ecosystem

This is evident in how the stock markets across the world react if there is a major crisis in a part of the world, and it has a ripple effect on the businesses elsewhere. With this background, consider the major crises in the past decade and half, to get the picture (Table 2.2).

All of the above crises had ripple effects on the global markets and have been the subjects of many research studies. The moral of the story is that in a highly connected world, expect disturbances at any time, even for reasons totally unrelated to your business, and this is just

Table 2.2   Global crisis list since the year 2000 (major ones only included)

Year	Region	Crisis
2000	USA	Telecom business crisis and dot-com bust
2001	USA	9/11 terrorist attacks, Turkish economic crisis
2002	South America	Banking crisis in Uruguay, Brazil, Venezuela
2003	Iraq	Invasion of Iraq by the USA
2004	Asia	Tsunami and Indian Ocean earthquake[a] with death toll of a quarter million people in 14 countries, and caused a loss of $20 billion
2007–2008	Global	Global financial crisis
2008	USA	Sub-prime mortgage crisis, automotive industry crisis
2008	Russia	Great economic recession
2009	Dubai	Financial crisis
2010	Europe	Eurozone crisis
2011	Japan	Tsunami
2014	Middle East	Oil crisis
2014	Russia	Financial crisis
2015	China	Stock market crash on June 12, 2015
2015	China	Devaluation of Yuan (CNY) currency for three consecutive times in August 2015
2016	UK	Brexit decision
2017	USA	Elections, tighter visa norms, and regulations
2017	North Korea	Missile and nuclear tests, leading to tensions

Source: https://en.wikipedia.org/wiki/List_of_economic_crises#21st_century
[a]https://en.wikipedia.org/wiki/2004_Indian_Ocean_earthquake_and_tsunami

because the world is much more connected today. There have been severe consequences, ranging from lack of availability of spare parts to downright closure of businesses and heavy job losses. New technologies have disrupted markets and have displaced large businesses. A typical example is what the Apple iPhone did to the mobile phone industry, and impacted established players like Nokia and Motorola. The smartphone also triggered explosive growth in the mobile app ecosystem and consequently the mobile data usage. This in turn disrupted other industries, for example, Uber impacted the taxi industry, social media impacted the Telecom sector's SMS messaging services, and so on.

It's important for you to build a resilient DNA for your business (and for yourself) from the start or at the earliest while constantly monitoring and upgrading your business and technology models.

2.3   Fragmentation Is Rife; Differentiation Is Key

Fragmentation and a plethora of choice is the new world order. This is due to specialization, granularity, regulations (e.g. privatization, de-regulation, etc.), and innovation/cross-fertilization fueled by ICT. The long tail of innovation will constantly spew out new choices, despite the addressable market being narrow for most of the choices. Even with consolidation at different levels and different parts of the ecosystem and industry happening from time to time, the sheer number of forces and the power unleashed by ICT and innovation will continue the disruption process repeatedly, at all levels and in all the parts. (Birudavolu 2015). Integration of any kind will carry unprecedented value. For doing this, one needs to have a clear idea of the big picture.

Amidst all this mayhem, formulating your unique value proposition clearly is key to survival and growth. Differentiating, specialization and focus are the new mantra. Hence organizations and individuals who track and know the big picture pertaining to their field, while simultaneously specializing in a specific area, will be the winners. Conversely, the losers will be those who do not specialize and have restricted awareness of the big picture. They will neither spot opportunities nor will they be able to capitalize on them.

Margins are one good indicator of success or failure. Low and continually falling margins indicate commoditization and risk. Automation may also eliminate the business. Whereas, healthy and high margins which hold out for longer periods indicate a good demand and strong potential for the business.

2.4   REGULATIONS: A CONSTANT CHALLENGE FOR GOVERNMENTS AND EVERYONE

With all the cataclysmic changes and paradigm shifts happening on a daily basis, the existing legal, regulatory, and industrial frameworks become outdated very fast. Both of the following are equally challenging:

- Upgrading regulations and formulating entirely new ones
- Compliance to regulations, implementing systems and processes fast that are in tune with the latest regulations and can withstand rigorous audits

Consider the following examples:

1. 3D printing is set to revolutionize manufacturing. Currently the taxes and duties for customs, sales, excise, and so on largely deal with physical goods moving across the borders of country or states. As the design of the goods grow more complex, the bulk of the value lies in the design, not the physical manufacturing of the finished product. Designs can be transmitted over the net, across borders, and used for 3D printing the goods anywhere. Thus this scheme can neatly bypass all the border controls and tax/regulatory compliance. This is not merely a revenue loss for the governments. Even guns can be 3D printed today. Thus new regulatory rules and controls need to be framed and implemented.

2. Net Neutrality has been the basic principle of the Internet. This means that all Internet traffic is to be treated alike. However if during peak hours, low-value, high-volume entertainment traffic is treated on par with premium business traffic or critical emergency traffic, then, in the long run, who will really pay for the Telecom's infrastructure, which will have to be rather massive to cater to the severe load during peak traffic?

3. Ease of banking and finance have also made money laundering and terror financing easy. The regulatory compliance will require a complex framework to be implemented to track the flow of finances, securities, and other matters relevant to law and order. This has to be done in a manner that does not overly burden all the parties in banking chain, but yet is able to maintain controls.

4. As machines become more intelligent and robots take on more work, replacing human beings, then how are contracts drawn up, and who is responsible for the mistakes and failures, and what are the legalities? This is critical where the stakes are high, such as software running the infrastructure and banks, and robots running operations where life is at stake. Examples range from industrial robots to autonomous driving cars and software bots at banks that take decisions on loans or cut financial deals. All these raise fundamental and deep philosophical questions about work, ownership, law, money, effort, collaboration, and so on. For example, should taxes be paid by (on behalf of) worker robots that have replaced humans?

5. Mobile telephony spectrum around the world is typically owned by the governments and is licensed to different players through auctions, and based on a number of criteria. All these are ridden with controversy. On the one hand mobile spectrum is valuable because anyone with a license can set up a lucrative business as a mobile phone operator. On the other hand, the auctions may lead to a heavy disparity as entrenched players with deep pockets can clobber the competition to grab most of the useful bandwidth. This would prevent new entrants with agile and efficient business models from entering into the market. The lack of competition would affect every aspect from innovation to pricing and monopoly, and this may not serve the public interest in the long run.

6. As the use of drones picks up, the aviation rules for drones will need to be drafted. The nature and types of drones and their uses will need to be defined and re-defined on an ongoing basis as technology and operations evolve. It should also address registration and tracking of drones, and also geo-fencing to keep out their operation in restricted areas.

Governments across the world are straining to cope with the changes both at a macro level and at a micro level. They need to set regulations that balance different factors: protection of public interests, defense of the nation, ensuring fair and free competition in the markets, tax structure, ensuring that the nation remains competitive on the global front, efficient use of resources and prevention of wastage, environment and sustainability, and so on. The challenge for governments is that the regulations need to be timely, amenable for implementation, and also forward-looking. The challenge for organizations is to implement and demonstrate compliance at all times. Knowledge bases and near complete automation will help on both sides.

2.5   SECURITY NIGHTMARES WITHOUT AN END

2.5.1   *The Magnitude of the Problem*

As software will eventually run everything, the vulnerabilities will also rise exponentially. This is a dangerous and widespread consequence of the digital tsunami. With software increasing its presence in every single field, and taking the central role, even the infrastructure, including public utilities like electricity, water, and transportation, are susceptible to hacking attacks. Cybercrime will cost businesses over US$2 trillion by 2019 (Juniper Research 2018).

To drive home the relevance and understand the stark magnitude of the problem, kindly read through a brief sample of *11 news items, all published just within the space of a few weeks.* The next two pages should leave you with no illusions about the terrible dangers that you and your business are exposed to:

1. *Cyber Attack on Infrastructure*:
   June 27, 2017: A cyber attack in Ukraine left *its electrical grid, national bank, and airport* crippled (Cyber Attacks 2017a).
2. *Cyber Attack on Banking*:
   June 29, 2017: European card fraud hits a record high of *1.8 billion Euros in 2016*, as per FICO (Cyber Attacks 2017b).
3. *Cyber Attack on Energy Sector Infrastructure*:
   July 7, 2017: Bloomberg reports that US officials have concluded that hackers working on behalf of foreign powers (primary suspect being Russia) have *recently breached at least a dozen US nuclear power sites* (Cyber Attacks 2017c).
4. *Cyber Vulnerability in Industrial Robots*:
   August 23, 2017: Cybersecurity research firm IOActive has demonstrated how robots can be hacked in ways that *could harm humans, by inflicting grave injuries* (Cyber Attacks 2017d).
5. *Cyber Attack on Telecoms and Citizens' Private Data*:
   August 23, 2017: Alibaba's UCWeb may face a ban in India for allegedly *leaking dozens of millions of users' mobile data to servers in China* (Cyber Attacks 2017e).

6. *Cyber Attack on Government and Employees:*

August 26, 2017: The FBI arrested a Chinese national responsible for a hack of the US Office of Personnel Management (OPM) and caused a data breach that compromised the information of *over 20 million* government employees (Cyber Attacks 2017f).

7. *Cyber Attack on Mobile Applications and Personal Devices:*

August 29, 2017: Google *removed about 300 apps* from its Play Store after they were found to *contain malware that hijacked the devices,* and could cause DDOS (distributed denial of service) attacks on other devices, that is, swamping other devices with data and rendering them unusable (Cyber Attacks 2017g).

8. *Cyber Attack on Digital Payment Systems:*

August 29, 2017: As per the US anti-money laundering software firm, Chainalysis, about *a quarter billion US dollars worth* of digital currency Ethereum has been lost due to cybercrime over the past year, *affecting over 30,000 people* (Cyber Attacks 2017h).

9. *Cyber Attack on Citizens' Data:*

August 30, 2017: In one of the world's largest data breaches ever, *over 711 million email addresses along with a number of passwords* were hacked open. The leak occurred after spammers hosted unsecured servers in the Netherlands, thereby allowing anyone to download information without requiring any credentials (Cyber Attacks 2017i).

10. *Cyber Attack on Medical Systems and Citizens' Health:*

August 31, 2017: The Food and Drug Administration (FDA) *recalled around 500,000 heart pacemakers* due to their risk of their cybersecurity getting hacked (Cyber Attacks 2017j).

11. *Cyber Attack on Citizens' Personal Data:*

September 2, 2017: Hackers of the Facebook-owned photo-sharing app Instagram sold users' email addresses and contact information *of over six million accounts* this week, at *US$10 per user* (Cyber Attacks 2017k).

12. *Cyber Attack on Industry and Citizens' Data:*

September 3, 2017: Over four million user records of Time Warner Cable have been exposed in a leak from Amazon's server, over the course of a month. *More than 600 GB of data* containing user's data have been compromised (Cyber Attacks 2017l).

13. *Cyber Attack on Consumers' Sensitive Data:*

September 9, 2017: In a breach that *affected 44% of the US population,* Equifax, a provider of consumer credit card reports, experienced criminal hacking on its website, thereby exposing *143 million US peoples' sensitive data*, including social security numbers, birth dates, addresses, credit card numbers, driver license numbers, and so on (Cyber Attacks 2017m).

September 17, 2017: Equifax hit with $70 billion lawsuit after leaking 143 million social security numbers (Cyber Attacks 2017o)

14. *Cyber Attack on Defense*:

September 8, 2017: WikiLeaks released hacked details for CIA's "Protego" missile system that was developed along with defense contractor Raytheon. The leak revealed 4 secret documents and 37 technical manuals related to the missile control system (Cyber Attacks 2017n).

2.5.2   Causes of Vulnerabilities

There are three causes of vulnerabilities. And any of these three causes or a combination thereof can render a software insecure against hacking attacks.

Bugs Hidden in the Software Released

These are flaws in the software and need fixing as and when they manifest. It can never be fully proven that any non-trivial software is completely free of bugs. Software validation and verification practices need to be very strong. Constant upgrade cycles need to be factored into the processes. There is an inherent cost.

To understand how insidious malware can exploit bugs, read about the Stuxnet work (Fruhlinger 2017), which even messed up uranium enrichment in Iran's nuclear reactors.

Human Error in Configuring and Customizing Software

Most software can be customized and configured in different ways, to suit specific needs, either organizational or individual. This is more so for enterprise software, because every organization has unique business needs. Many omissions and commissions result during this process when done by different development and operations teams over the months and years. These invariably open up a host of security vulnerabilities, among other problems.

Obsolescence
In a rapidly evolving world, even the best of software becomes outdated quickly, and will need to be upgraded frequently to function in the larger ecosystem. An out-of-step software can not only result in mismatches of various kinds but also result in many security vulnerabilities. In fact one huge challenge in a full stack of software products is that different components need to be constantly upgraded and all of them together need to function well. The so-called "zero day attacks" are a good example of this. In these, the moment a software vulnerability is discovered in a layer of software, if the software patch is not applied immediately, hackers around the world exploit the opening on day zero, that is, in the initial period, when the patch is not yet well known to the public, or people have delayed applying it. Millions of sites are thus open to attack because they are running an outdated software from the security point of view.

It's important to note that the time when the outdated software was built and released, the current vulnerability may never have been considered a bug or a defect at all.

This situation is similar to replacing the old padlocks for all the doors in a large building to install strongly secure modern electronic locks, but somehow forgetting to upgrade one entry door in the building's rear. The entire building stands compromised now, because passkeys for old padlocks are easily available today, as compared to decades ago when the padlocks were considered secure. And imagine the scenario if there are thousands of people attempting to break into the building every hour, using every conceivable means. This is analogous to the systems online on the internet.

A Quiz Question
If you were to set up a server on the Internet today with a weak password like "abcd1234", how long would it take before the server is hacked into and your data destroyed, stolen, encrypted off, and rendered useless due to a ransomware demand? Your six choices for the answer to the question are:

(a) *10 minutes* (b) *6 hours* (c) *1 week* (d) *1 month* (e) *6 months*

If you've answered (a), you're probably quite right. There are thousands of malicious bots running loose on the Internet that keep attempting to break into sites using every means possible. They're completely automated and highly efficient. Note that *even* if you were to pay the demanded ransom (presumably in bitcoins), it is extremely unlikely that you'd get back your data.

In the Dark Net there are sophisticated Do-It-Yourself programs using which you can build your own customized malware. The program will let you choose a combination of your choice of a vehicle and its payload. For example, you could choose your vehicle for delivering a malware, for example, an email attachment, a bot, an infected pendrive, and so on, and then choose a payload, that is, the effect the malware would have, such as deleting the target computer's files, or secretly copying and transmitting the files, or encrypting files and demanding ransomware, or spying on the activities on the computer, and so on. Anyone can thus create a custom malware to order, and pay in bitcoins for the creation!

It is a fair assumption that most of the infrastructure and systems *are already hacked*, and are merely waiting for the opportune moment to unleash havoc or have already taken and sold the information.

2.6   RISING INEQUALITY

The surest impact of the digital tsunami is that on the global workforce. Specialization and automation are the twin forces responsible for this disruption. Automation has a direct impact on jobs. Deutsche Bank CEO, John Cryan, has said on September, 7, 2017, that a large number of his staff will be retrenched due to automation. He adds: "In our bank we have people doing work like robots. Tomorrow we will have robots behaving like people." Automation will invade every area, every domain, every sector, and sub-sector. It will sweep through traditional and non-traditional areas. For example, all the arts in the future will have a large component of automation right from the inception through processes of creation and production/output. Generative neural networks are a cutting-edge example. In most areas, Artificial Intelligence will completely outpace human intelligence and replace the latter.

Specialization helps in differentiation and innovation. It guards against commoditization, severe competition, and low margins. Specialization is also a moving target, for example, selling online was once a specialization, but is no longer so, because the entry barriers are quite low. The same logic holds for individuals also. Hence the bulk of the current workforce, which has no specialized skills, will be severely hit by this new paradigm shift. Specialization also leads to fragmentation. There are fewer general physicians today than before, and they don't make as much as the specialists. The "general" workforce will suffer. Inequality will be rampant.

The workforce population generally falls into three categories (actually a spectrum):

1. Those with specialized skills, for example, cutting-edge programming languages, domain expertise in prime areas such as data sciences, biochemistry, engineering, medical, finance, and so on
2. Those specialized in skills that are already commoditized or even obsolete, for example, Microsoft Office skills, outdated programming languages, data entry, very common "management" skills acquired through experience, and so on
3. Those with no specialized skills or low-skilled, for example, most blue collar jobs, taxi drivers, ticketing clerks, and so on

The first category will obviously dominate the rest of the population and will drive the future. This will be the category where most of the high-valued jobs of the future will be created. The second category will survive in the market, if they constantly upgrade their skills and are willing to work in new fields, relocate, work remotely, and become contract workers. It is the third category of workers who are at maximum risk of losing their jobs altogether or getting retrenched altogether as new technologies and automation eat away at their employment potential, for example, new technology of shoe-making removed most of the cobblers' jobs; ready-made clothes did away with most tailors; email and messaging did away with paperwork, printing, clerks, telegraph, fax, and most of postal workforce; modern construction equipment and the standard blocks for building material reduce laborers.

Due to focus and specialization, organizations will outsource all non-core activities. All non-core work will be viewed as overheads, meant to be minimized. This means a rise in automation and in contract workforce. The latter will be under constant and severe cost pressures, as against the core specialist employees, who will earn premium pay and benefits. Companies may not provide the contract employees with benefits such as medical insurance, training, education, leaves, maternity benefits, perquisites, loans, and stocks. They will be left to fend for their own. Living standards will become expensive for them. Jobs and wages will fall for the second and third workforce categories, and severe inequality will set in the societies for a long period of time. Unfortunately these constitute a large portion of the populace. The population will be divided into digital "haves" and "have-nots" for at least one generation more, after which the gap will steadily reduce, and the entire population will be ushered into a new era with its own challenges and solutions. Urgent policy measures and proper investments will greatly help in reducing the divide.

2.7   WHERE ARE THE OPPORTUNITIES?

To find sizeable opportunities, look for sizeable problems and significant pain points. It is easiest to start by looking at inefficiencies in the ecosystem. And next, look at the unmet needs, especially in the long tail. The opportunities are discussed in greater detail in the upcoming chapters. The future poses new challenges, which will need new solutions. The size of the solution need not match the size of the problem. Asymmetry is in fact preferred. A large problem may have a simple solution. For example, if there are two (unrelated) problems A and B, concerned with eliminating wastage worth $1 billion and $20 million respectively. The solutions for A and B need not incur a proportional amount of expenditure; they could have solutions for only $1000 and $5000, respectively. In true innovation, the magnitude of the problem may have no bearing on the type, extent, and value of the resources needed for the solution.

A few examples:

- *Banking*: Connecting to a million customers today may be achieved with an intelligent chatbot, such as State Bank of India's intelligent chatbot SIA (www.sbi.co.in/sia/index.html).
- *Precision Farming*: In crop irrigation, optimizing water flow and preventing millions of liters of water wastage (and wastage of electricity) may be accomplished with a simple IoT solution that measures the soil moisture with a few sensors that will trigger an on/off switch to the pump, all monitored by a low-cost cloud solution. Example: FlyBird Innovations (www.flybirdinnovations.com).
- *Education*: Training a million students and tracking/helping their progress in a graded manner are made possible through an online course that uses AI to measure individual progress through quizzes and guides them to the relevant material. Example: Embibe (www.embibe.com).

It is all too important to first understand the problems in a great degree of depth, especially when faced with all the challenges outlined earlier in this chapter. The new era favors low footprint and sustainable solutions. New business models need to be experimented with fast and improved through iterations. One cannot sink in significant investment into any version, without having prototyped and tested many variants. Even so, it is better to make that significant investment in a suitable underlying platform and framework that will support many different and collaborative business models, rather than betting everything on a single business model. Because with changing market conditions and new innovations, the business model itself should withstand changes and tweaks, or even be replaced by another.

Formulating solutions will need a thorough approach. This includes innovation, culture, strategy, tools and frameworks, collaboration, technology, and so on. All these are discussed well in the remaining chapters.

To sustain leadership in the markets over time, organizations and individuals need to seek out meaningful, purposeful opportunities that they can capitalize on. It is good to use a framework like Ikigai (Miralles and Garcia 2016) (Nash 2018). Startups are typically started by very few people (typically one or two). Hence the principles of Ikigai work well there also. As an organization grows, it develops an identity and a DNA. An innovative product launched should also resonate with the DNA of the organization. The Japanese concept of Ikigai suggests that one must find a profession or work that is at the junction of these four circles:

1. What you *Love*
2. What you are *Good* at
3. What the world *Needs*
4. What you can be *Paid* for

Ikigai generally holds for all individuals and organizations. There are exceptions like NGOs or non-profit organizations that don't have a primary goal of making money, or companies that specialize in sustainability and rely for funds on CSR (corporate social responsibility) funds, and these organizations can afford to compromise in some areas if they have their overall model worked out thoroughly. But it is good for the other organizations to follow Ikigai.

If any one or more of the four circles are weakly represented or left out, then it is difficult for the organization to thrive in the (competitive) market for long. For example, it could run into problems like:

- Circle #1 Missing: Work may become drudgery if it is taken up merely for money; hence the enthusiasm cannot become contagious; the teams won't get motivated to do an outstanding job to remain competitive.
- Circle #2 Missing: The organization is unable to build it despite the management's passion because it has no wherewithal to execute it.
- Circle #3 Missing: The organization may build something interesting but has no market.
- Circle #4 Missing: Weak revenue model leading to unsustainability.

Missing combinations of these lead to complications from which it is difficult for the organization to extricate itself.

On the other hand, if an organization positions itself at the junction and remains focused, it becomes a deadly force to reckon with in the market, due to its natural advantages. Many a time, it takes a few iterations to figure out your unique circles, that is, what belongs in your circles. And that should be considered quite normal.

2.8   Conclusion

There is a plethora of disruptions possible in the current and future markets. Disruptions can be triggered by a variety of factors such as economic shakeups, new technologies, new business models, changes in regulations, competitive collaborations, and so on.

A complete change in mindset is needed. It's essential to shift from a mindset of security and stability to that of resilience. The leadership needs to move away from setting goals and strategies that revolve around establishing a secure and safe business, to those that help build a business which is opportunistic but resilient to the inevitable shocks and disruptions. This requires developing organizations that are very focused, innovative, risk-taking, experimental, which co-create, and have imbibed intrapreneurship and ambidextrous thinking in their DNA. The organization needs to grow or re-structure itself for exploiting market opportunities that are at the junction of the four circles, consisting of their passion, strengths, available/new markets, and those which are lucrative.

Technology brings both opportunities and challenges. It is the wise vision of the leadership which can drive the organization through ICT Tsunami. First, an organization needs to understand the economic value of the ICT which can be nurtured through innovation and strategy development by incorporating ICT tools. A company requires an idea as to how to upgrade the production or service delivery system through collaborating with partners and even with competitors. As mentioned above, fragmentation is rife and differentiation is the key for success. The economic value of ICT tools depends on fragmentation strategy. The whole thing is further dependent on regulatory environment and level of competition in the economy. How far regulation drives the adoption of new technology and how the nature of competition shapes the business environment, the innovation strategy will be shaped accordingly. Finally, the ICT Tsunami has brought the challenges in the form of a data security nightmare and fast obsolescence of technology. On the other side, technology provides enough opportunity to exploit the economies of scale and create new market for the innovated products and services. The figure below provides a snapshot of the above discussion (Fig. 2.1).

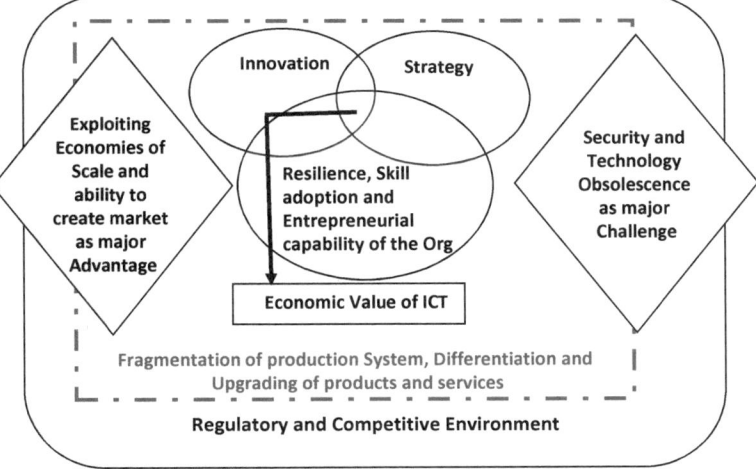

Fig. 2.1   ICT strategy: opportunities and challenges

> **The Takeaway Box: ICT and Innovation Are Reshaping the World—Including Your Business and Your Career**
>
> - Use the checklists given in this chapter to help you migrate to the new order quickly.
> - In this uncertain and insecure world of crisis, fragmentation, changing regulations, and hyper-competition:
>   *specialization, innovation, collaboration, and integration are the keys to survival*
> - It takes a holistic, multi-pronged approach to survive and thrive in business and in your career through the storms. You may need to pivot your business/yourself to reach success.
> - Take the material in this chapter and the remainder of this book seriously!

REFERENCES

Airbnb. (2018). Retrieved from https://en.wikipedia.org/wiki/Airbnb.

BioCarbon Engineering. (2018). Retrieved from www.biocarbonengineering.com.

Birudavolu, Sriram (2015): "Open Innovation in ICT: An Empirical Assesment of Global Telecommunications Services"; Unpublished Thesis submitted for PhD at Indian Institute of Foreign Trade, New Delhi

Biesheuvel, T. (2017, June). How Just 14 People Make 500,000 Tons of Steel a Year in Austria. *Bloomberg*. Retrieved from www.bloomberg.com/news/articles/2017-06-21/how-just-14-people-make-500-000-tons-of-steel-a-year-in-austria.

Capgemini Consulting. (2015). Retrieved from https://www.capgemini-consulting.com/resource-file-access/resource/pdf/digital_disruption_1.pdf.

Cyber Attacks. (2017a). *Ukraine Attacks*. Retrieved from http://www.independent.co.uk/news/world/europe/ukraine-cyber-attack-hackers-national-bank-state-power-company-airport-rozenko-pavlo-cabinet-a7810471.html.

Cyber Attacks. (2017b). Retrieved from http://www.fico.com/en/blogs/analytics-optimization/the-story-behind-europes-e1-8-billion-card-fraud-problem/.

Cyber Attacks. (2017c). Retrieved from https://www.bloomberg.com/news/articles/2017-07-07/russians-are-said-to-be-suspects-in-hacks-involving-nuclear-site.

Cyber Attacks. (2017d). Retrieved from http://www.newsweek.com/hacked-killer-robots-serious-threat-humans-property-562153.

Cyber Attacks. (2017e). Retrieved from https://www.vccircle.com/alibabas-ucweb-under-govt-scanner-over-data-leak-may-face-ban/.

Cyber Attacks. (2017f). Retrieved from https://www.reuters.com/article/us-usa-cyber-opm/chinese-national-arrested-in-los-angeles-on-u-s-hacking-charge-idUSKCN1B42RM.

Cyber Attacks. (2017g). Retrieved from https://www.theverge.com/2017/8/29/16219426/google-removes-apps-play-store-hijack-phones-ddos-attacks.

Cyber Attacks. (2017h). Retrieved from https://www.bloomberg.com/news/articles/2017-08-24/cyber-criminals-extracting-a-heavy-toll-from-ethereum-advocates.

Cyber Attacks. (2017i). Retrieved from http://www.zdnet.com/article/onliner-spambot-largest-ever-malware-campaign-millions/.

Cyber Attacks. (2017j). Retrieved from https://www.theguardian.com/technology/2017/aug/31/hacking-risk-recall-pacemakers-patient-death-fears-fda-firmware-update.

Cyber Attacks. (2017k). Retrieved from http://www.express.co.uk/life-style/science-technology/849158/Instagram-login-hack-download-user-e-mails-Selena-Gomez-Emma-Watson-Emilia-Clarke.

Cyber Attacks. (2017l). Retrieved from https://www.cnbc.com/2017/09/01/around-4-million-time-warner-personal-records-exposed-in-data-leak.html.

Cyber Attacks. (2017m). Retrieved from https://arstechnica.com/information-technology/2017/09/equifax-website-hack-exposes-data-for-143-million-us-consumers/.

Cyber Attacks. (2017n). Retrieved from www.wikileaks.ch.

Cyber Attacks. (2017o). Retrieved from http://www.zerohedge.com/news/2017-09-08/equifax-hit-70-billion-lawsuit-after-leaking-143-million-social-security-numbers.

Forbes. (2018, February 22). *Breaking Down Uber's Valuation: An Interactive Analysis.* Retrieved from www.forbes.com/sites/greatspeculations/2018/02/22/breaking-down-ubers-valuation-an-interactive-analysis.

Fossbytes. (2017, December). *Google's AI Creates Its Own AI That Beats the Performance of Other Human-Made Models.* Retrieved from https://fossbytes.com/google-ai-automl-nasnet-performance/.

Fruhlinger, J. (2017, August). What Is Stuxnet, Who Created It and How Does It Work?, *CSO.* Retrieved from https://www.csoonline.com/article/3218104/malware/what-is-stuxnet-who-created-it-and-how-does-it-work.html.

Innosight, 2018, Corporate Longevity Forecast: Creative Destruction is Accelerating, https://www.innosight.com/insight/creative-destruction/

Juniper Research. (2018). *Cybercrime Will Cost Businesses over $2 Trillion by 2019.* Retrieved from https://www.juniperresearch.com/press/press-releases/cybercrime-cost-businesses-over-2trillion.

Keller, M. (2016, September). *These Engineers 3D Printed a Mini Jet Engine, Then Took It to 33,000 RPM.* Retrieved from www.ge.com/reports/post/118394013625/these-engineers-3d-printed-a-mini-jet-engine-then-.

Kira. (2015). *ORNL Unveils Integrated 3D Printed House and Car That Produce and Share Clean Energy*. Retrieved from www.3ders.org/articles/20150923-ornl-unveils-integrated-3d-printed-house-and-car-that-produce-and-share-clean-energy.html.

Livemint. (2017, September 17). Retrieved from http://www.livemint.com/Industry/2EAq3qUEnUvgemPy6nX4SL/ExCiti-CEO-Vikram-Pandit-says-30-of-bank-jobs-at-risk-from.html?utm_source=inshorts&utm_medium=referral&utm_campaign=fullarticle.

McFarland, M. (2016, October). *Amazon Only Needs a Minute of Human Labor to Ship Your Next Package*. Retrieved from http://money.cnn.com/2016/10/06/technology/amazon-warehouse-robots/index.html.

Microsoft. (2017, August). *Microsoft Researchers Achieve New Conversational Speech Recognition Milestone*. Retrieved from www.microsoft.com/en-us/research/blog/microsoft-researchers-achieve-new-conversational-speech-recognition-milestone.

Miralles, F., & Garcia, H. (2016). *Ikigai: The Japanese Secret to a Long and Happy Life*. Retrieved from https://books.google.com.

Mobile-only Banks. (2018). Retrieved from http://mobileonlybank.com/.

Mourdoukoutas, P. (2017). Alibaba Beats Amazon. *Forbes*. Retrieved from www.forbes.com/sites/panosmourdoukoutas/2017/08/22/alibaba-beats-amazon.

Nash, A. (2018). *Ikigai: How to Find Professional Success*. Retrieved from https://adamnash.blog/2018/02/06/ikigai-how-to-find-professional-success/.

Netflix. (2018). Retrieved from https://en.wikipedia.org/wiki/Netflix.

Nike. (2018). Retrieved from https://en.wikipedia.org/wiki/Nike,_Inc.

Paris, C. (2017, May). Norway Takes Lead in Race to Build Autonomous Cargo Ships. *Wall Street Journal*. Retrieved from www.wsj.com/articles/norway-takes-lead-in-race-to-build-autonomous-cargo-ships-1500721202.

Perry, M. J. (2017, October 20). *Fortune 500 Firms 1955 v. 2017: Only 60 Remain, Thanks to the Creative Destruction That Fuels Economic Prosperity*, American Enterprise Institute. Retrieved from http://www.aei.org/publication/fortune-500-firms-1955-v-2017-only-12-remain-thanks-to-the-creative-destruction-that-fuels-economic-prosperity/.

Rolls-Royce. (2015). *Rolls-Royce Trent XWB-97 Completes First Test Flight*. Retrieved from www.rolls-royce.com/media/press-releases/yr-2015/pr-06-11-2015-rolls-royce-trent-xwb-97-completes-first-test-flight.aspx.

Saunders, S. (2017, October). *GE Aviation Completes Successful Tests of FATE Engine and T901 Engine Prototype, Both Featuring 3D Printed Components*. Retrieved from https://3dprint.com/190427/ge-aviation-engine-test.

Simonite, T. (2018, February 28). Using AI to Help Stroke Victims When 'Time Is Brain'. *Wired Magazine*. Retrieved from www.wired.com/story/using-ai-to-help-stroke-victims-when-time-is-brain.

Weiner, S. (2016, September). *Facebook's Arctic Data Center is Eeerily Beautiful.* Retrieved from www.popularmechanics.com/technology/g2792/photos-of-facebooks-swedish-data-center-are-eerily-beautiful/.

Wikipedia, Alibaba. (2018). Retrieved from https://en.wikipedia.org/wiki/Alibaba_Group.

Wikipedia, Amazon. (2018). Retrieved from https://en.wikipedia.org/wiki/Amazon_(company).

Xia, Z. (2016). *JD.com Builds World's First Fully Automated Warehouse in Shanghai, Yi Cai Global.* Retrieved from https://yicaiglobal.com/news/jdcom-builds-world%E2%80%99s-first-fully-automated-warehouse-shanghai.

## CHAPTER 3

Every Organization Is a Complex ICT System

Architecture does not create extraordinary organizations by collecting extraordinary people. It does so by enabling very ordinary people to perform in extraordinary ways.
—*John Kay (1995) Elements of Corporate Success*

3.1 Introduction

We can take the working definition of an organization as:

An organized group of people with a particular purpose, such as a business or government department.

The theme of this chapter is that *any organization may be fundamentally viewed as a complex information processing engine.*

An organization has many facets, such as vision, goals, people, structure, process, collaboration, activities, roles, responsibilities, relationships, and environments. Each of these basically constitutes *information and communication*, shared in the right context to various stakeholder groups. And, the context itself, when analyzed, is meta-information, that is, essentially information again, albeit of a higher order.

If this is hard to imagine, read through the following cases, ranging from the simple to the sophisticated:

© The Author(s) 2019
S. Birudavolu, B. Nag, *Business Innovation and ICT Strategies*,
https://doi.org/10.1007/978-981-13-1675-3_3

1. Aerospace and Manufacturing

   Take a large company like Airbus or Boeing. A Boeing 747-400 has six million parts that are designed, manufactured, assembled, and tested. An Airbus A380 has an approximate 4 million parts, produced by 1500 companies from 30 countries around the world. The research and design process all results in design specification of each component, and subsequently the integration and testing, *all of which is information*. When it is given out to vendors for manufacturing, contracts are drawn up, which is also information. The quality checks, data, and approvals are all *information* again. The specifications and process at every step need to be so thoroughly defined at a fine-grained level, so as to leave no ambiguity, and to even enable software and robots to perform the tasks. Thus, the final airplane is nothing but a final physical representation of a sea of *information* behind it.

2. Logistics

   Consider an operation in which steel bars are physically moved from point A to point B, and a person keeps tab on the quantity transported. When the organization is IT enabled, an ERP system may tally the quantity, quality, value, batch number of steel moved, the date, time and the duration it took for the operation, and the resources involved in carrying out the operation. If the operation gets automated or outsourced, the information about the operation *is the only part that remains and is important* to the organization. The operation itself may be eliminated due to an overall optimization exercise. Then the physical operation will stand canceled, because of re-engineering resulting from the optimization carried out based on the information model and using the operations data gathered over time.

   Thus, generally the information and communication eventually drive the physical operations *and not the other way around* Operations are used to gather data to validate and improve the information model.

3. Pharma

   A pharma company probably takes a decade to bring out a drug molecule into the market. Its research and development labs' experiments, the drug selection, the drug approval process by the regulatory authority, the outsourcing of the drug manufacturing, the quality checks, the eventual rollout into the market, *all of these constitute* information and communication.

4. Abstract Concepts

   Many abstract concepts can be distilled into concrete parameters, measured directly or indirectly, to render them suitable for use. Examples include a tourist's experience of a trip, opinion polls, the likeability of a specific perfume, the flavor of a brand of tea, the taste of a biscuit, perceptions about an airline or a hotel, appeal of a luxury car, power of a relationship, and so on. This is not mere theorizing, but billions of dollars are invested in finding and using these concrete parameters, which drive the marketing strategy and operations of companies engaged in these businesses.

5. Sustainable Agriculture

   The Netherlands is a small country unsuited for large-scale agriculture because it is only about a 1000 km away from the Arctic Circle. But it is the world's second largest exporter of food in dollar terms, the largest being the USA, which has a land area 270 times larger than the Netherlands (Viviano 2016). The Netherlands is the topmost global exporter of potatoes and onions, is the second largest exporter of vegetables, and generates one third of the global trade in vegetable seeds. And it does all this without the use of chemical pesticides, GM crops, or antibiotics (for the livestock). It has accomplished all this through the following means:

   (a) Large greenhouses, with climate-controlled farms, constitute 80% of the cultivated land—control every parameter carefully through Information Technology, thereby eliminating human intervention.

   (b) Use of driverless tractors, drones, and sensors that measure everything—water readings, soil chemistry, nutrients, measuring growth of every individual plant.

   (c) Reduce dependence on water by 90%. For example, each kilogram of tomatoes requires less than 4 gallons of water compared to 16 gallons needed in an open-field cultivation.

   (d) Use of LED lighting for 24-hour cultivation in controlled environments, entirely automated, and to produce micro-algae for supplying proteins and lipids for the food chain.

   (e) Produce fish and vegetables in a self-sustaining loop—fish produce waste that is natural fertilizer for the plants, which in turn clean up the water for the fish.

(f) Controlled cultivation of predatory microorganisms that destroy the pests.
(g) Molecular breeding of seeds, without GM (genetic modification).

All this is accomplished through deep research and collaboration. WUR (Wageningen University & Research) is at the center of the entire Food Valley, a cluster of agri-startups.

The Netherlands shares its practices to different developing countries around the world like China, India, Indonesia, and so on in the interests of solving the world's food/hunger problem and promoting sustainable practices. Hence knowledge is the main export.

3.2   Every Business Is Now a Digital Business

Countless articles have been published on why software is eating the world. Hence it is vital to look at every business and operation thoroughly through the digital lens, before your competitors do. Even if you are a monopoly or dominate a niche area, your business will fall rapidly out of sync with the environment and ecosystem in which you function if there is no digital play. This will give rise to competition that will match the market needs efficiently, and your business will get pushed downstream, into being commoditized and perhaps rendered obsolete.

Consider the following simple, but instructive, example:

A small restaurant which does nothing but cook food and serve customers that come to its doorstep could still face steep competition from digital aggregator food service vendors, who can service those very customers in the comfort of the latter's homes at the click of a mobile app. The digital vendors could offer very competitive prices despite the door delivery because they could source from low-cost suppliers who prepare food in bulk in a no-frills, inexpensive location. Through the digital world, companies can connect to customers much faster and more efficiently, and traditional restaurants lack that capability. Digitization also helps in streamlining the backend, the supply chain, input optimization, and process simplification. Payments can be done through mobile money and payment gateways seamlessly integrated with the mobile and web applications.

The restaurant will still need digitization to understand and manage its own operations thoroughly well, such as daily/seasonal fluctuations in business at an intricate level, costs, storage, procurement of raw foodstuff, reduce wastage, manpower, cooking and serving processes, taking orders, offering menu suggestions to customers, and so on. The restaurant can also become a supplier to the digital aggregators, especially during lean periods, or even host part of its services online, and take retail and bulk orders directly from customers.

Examples of online retail food chains, including aggregators, are TinMen (www.tinmen.in), Swiggy (www.swiggy.com), and InnerChef (www.innerchef.com).

Digitalization, of course, requires that all the processes of the restaurant be thoroughly analyzed and simplified and streamlined first. It is futile to automate chaos and inefficiency. Business processes need to be re-engineered.

Digital marketing will help boost the restaurant's popularity, in many ways, such as showcasing its menu, ease of people finding the restaurant through web search, giving driving directions on a map to the restaurant's location, promotion through publishing positive feedback online, improving strategy and operations by working on negative feedback, pre-bookings through mobile app, discount vouchers, loyalty plans, and so on.

Haptik (www.haptik.ai) is a company that is a chatbot platform. It has an AI-powered conversational interface that uses deep learning and Machine Learning that can interface with human beings and drive engaging conversations, mimicking a real human agent. It helps companies in lead generation, customer support, customer utility, and sales enablement. For example, this is useful in setting up a 24/7 customer support channel.

As is evident, digitalization also enables both deep and extensive automation of operations. A platform approach on the cloud would enable the restaurant to seamlessly integrate its procurement, operations, finances, and link up with other organizations, such as suppliers, supply chain logistics, loyalty points vendors, search engines, travel sites, food and cuisine sites, aggregators, events, and so on.

Analytics, Artificial Intelligence, and Machine Learning would enable deeper and interesting patterns to be discovered automatically, thereby enabling superior predictions and decisions in every area including demand forecasting, pricing, customer wait-time in queues, costing, profitability of menu-items, deal/package offers, stocking, customer profiles, competitors, customer satisfaction, recruitment, payroll, outsourcing, equipment and raw material purchase decisions, and so on. Every aspect of the business and operations can be analyzed very deeply and appropriate decisions can be taken. For example, data will allow us to analyze the cooking process, for finding what dishes take how much time, and what mix-and-match combinations work best for speeding up deliveries. It may lead to timely and right decisions regarding upgrading stoves or purchasing new ones, maintenance contracts, space management, hiring cooks, and so on.

Once the restaurant has digitalized and perfected its strategy and operations into a lean and automated business, it could, should it so desire, look at expanding its operations, organically or through franchising, or by integration with other restaurants chains and major hotels. Digitalization and automation will empower the restaurant to engage in a wide variety of collaborations and partnerships and open innovation. The restaurant can collaborate to seamlessly provide services to the aggregators such as those mentioned in the earlier section.

There is a social angle also. The restaurants (especially the high-end ones) can dispose of their unused, surplus food to orphanages, schools, and hostels, instead of wasting it. With a digital platform, this can be better managed.

Hence it is largely true that every business is now a digital business, regardless of whichever sector it may be in.

3.3   Digital Revolution and Economics of Business Development

As mentioned above every organization is a complex ICT system and every business is a digital business, the finer balance for an organization depends on how it innovates new products and services and how it upgrades old products using the ICT tools. One of the important assumptions here is that the number of products and services that embraces digital technologies is increasing over time. This has accentuated due to intense competition. Hence, whenever, any company thinks of a new product it has to articulate ICT tools for delivering the same. For example, take any university department that wants to launch a new product (as new course). It can launch it through classroom mode or through the online mode. In the case of classroom mode, it has to bring digital technology such as computer, overhead projection, Internet, online material, new software, and so on. In the case of online teaching, it requires sophisticated Internet-based technology and entire online-based course material as students are located in different cities. So, technology creates a pressure on the system for product diversification. If the progress of technology is fast and disruptive, an innovative strategy can provide new value which may last for a longer duration. Hence, any company can sustain the technology pressure for a longer period if it simultaneously innovates. The following diagram explains this. With the introduction of fast technological progress (disruption or Tech Tsunami), the number of products and services adopting new technology goes up substantially, and if a company adopts new technology, they can derive higher value and thereby can have more control in the market. As a result, value curve moves up as described in the diagram. Hence, companies need to embrace new technology to remain in the business and needs to see every business as an application of new technology to derive higher and sustainable value. As mentioned earlier that every business generates lots of data, digital technology helps companies to use these data intelligently to change the product profile, get more consumer insights, create new products and services as per the demand, streamline the supply chain, and hence generate huge amount of new "value" which may be precious for sustenance (Fig. 3.1) (Birudavolu 2015).

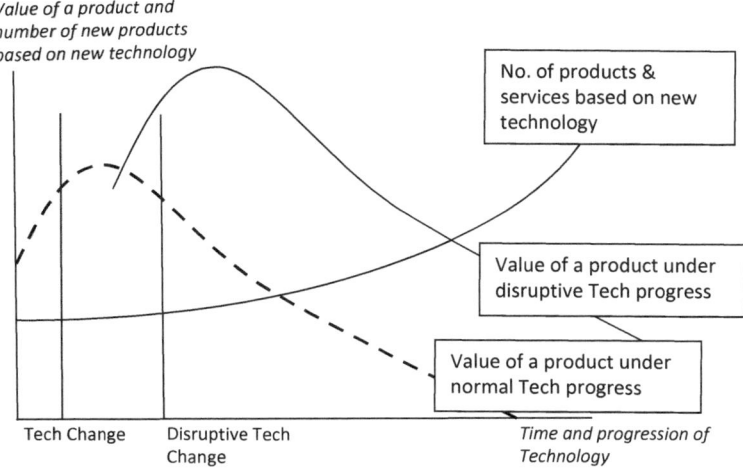

Fig. 3.1 Technological change and sustainability with new products (Source: Authors' own creation based on various sources and ideas discussed in Roberts Edward B (2002). Innovation: Driving Product, Process, and Market Change, *MIT Sloan Management Review*, published by Jossey-Bass, San Francisco, and Afuah Allan (2009), *Strategic Innovation: New Game Strategies for Competitive Advantage*, Routledge, New York)

3.4   Information Valuation

The cliché "Data is the new Oil" has truth in it. Next to its people, an organization's data is among its most valuable asset. People change roles and departments or leave the organization. New people may join in. But data remains a permanent repository, as a faithful record and witness to all that the organization is doing. It is important to value and valuate data. Gartner suggests (Laney 2017) that there are six formal information valuation models:

The Foundational Measures (Organizational Process Perspective):

1. Intrinsic Value: correctness, completeness, and exclusiveness of the data.
2. Business Value: relevance and goodness for a specific purpose.
3. Performance Value: effect of this data on the key business drivers.

The Financial Measures (Economic Perspective):

4. Cost Value: damage cost/value if we were to lose this data.
5. Market Value: value measured if we sold this data in the market.
6. Economic Value: this data's contribution value to our top/bottom line.

As an organization matures and becomes more information-centric, more sophistication can be added.

3.5   The New Organization

To thrive in the new world and to be able to leverage new paradigms requires a totally new kind of organization. The current organization structures are built for creating products/services that are monolithic and inflexible. The model is rapidly becoming obsolete, and hence needs urgent replacing.

Organizations face a severe risk due to:

- Technological obsolescence
- Market obsolescence
- Competitor's moves (need to have ability to react very fast to these)
- The need to exploit new opportunities, in the market, that need to be competitive

Current organization models hit a limit soon because they cannot grow themselves in time to meet these challenges. The reasons are:

- The entire structure is incentivized, trained, institutionalized, and hardened for years in efficiently delivering a fixed set of product lines with limited variations.
- The organization is built to produce chunky/coarse products and services that execute a limited range of functionality. They cannot be segmented in a fine-grained manner to be mixed and matched in different ways or combined with external products/services to produce new varieties that suit the changing market needs.
- It is too complicated, time-consuming, and difficult to introduce major changes across all the silos and layers of the hierarchy of the traditional organization; it needs massive coordination, cutting across turfs (political, business, financial, organizational teams).
- The business logic and flows are siloed, fragmented, and embedded deeply across the organization; hence there is no single ownership for delivery of business functionality.

Thus companies take a path of inorganic growth, and engage in mergers and acquisitions. Or seek to get acquired. These are fraught with challenges, if there are mismatches in culture or in merging/integrating the product lines and the relevant teams. There are financial pressures and challenges too. The merged organization has to justify and recover the cost of acquisition/merger while taking the organization through the changes, continuing to run operations at top speed in a business-as-usual mode, and keeping customers happy. Retaining top talent is another major challenge.

The new organization on the other hand is built differently. It develops fine-grained components and microservices that:

- Are constantly evolving
- Can be combined in different ways to product new kinds of products and services
- Provide for integration with external products/services, thereby enabling open innovation
- Are highly automated

The integration, both external and internal, is accomplished through a platform approach. Each team focuses on producing and perfecting one component/microservice. Hence there are small, distributed, specialist teams that carry a high sense of ownership. There is a lot of innovation at play everywhere in the new organization, due to the endless possibilities for delivering different types of services, with new ones coming up every day, and in keeping abreast of the industry.

The architecture and design are quite simplified because of the high ownership everywhere in the organization (not just at the top/center), the clean interfaces, and inherently scalable design. The Kaizen principle of continuous evolution through incremental progress can be perfectly applied here. The communication across teams and organizations improves due to simplicity. The signal-to-noise ratio improves drastically because of low latency, lack of hierarchies, and clean interfaces.

In the new organization, development and operations are intertwined into DevOps, an agile, seamless, and efficient process. It is also highly amenable to automation.

In the new organization, hiring and talent retention are the most important activities. There is a war for talent. Human resources are considered more precious than financial resources or other organizational resources. In a lean organization, there is a clear vision, strong strategy, and highly efficient, automated operations. It has really a smart and dynamic workforce in every area. Hiring happens throughout the year, depending on when the right talent is available, not when finances are available. A large percentage of the workforce could be on contract, not permanent employees. It will be a "gig economy". The entire industry will work largely on a "fee-for-delivered-value" basis.

Another important factor is that much of the workforce will be distributed, not working out of headquarters or a few offices. With increasingly affordable advanced tools like Cisco Webex, Cisco Spark, Zoho Meeting, and Skype, unified communication will become ubiquitous.

Working remotely or from home will be an accepted norm. The size of offices will be shrunk to accommodate a fraction of the workforce. Flexible seating will be the norm. These will help in slicing down real-estate costs and other usual overheads. It doesn't really matter where you are, when you work, or the kind of tools you use; only the framework, the quantum of work, and the outcomes matter.

As described in this chapter, ICT makes all this possible, like never before.

The following ingredients are vital for building a top-class organization:

(a) *Continuous Learning*—It is important to develop a learning organization (Senge 1990). Training and development are of the highest priority in a world that is rapidly evolving. A surprising development will be that the training will be done not merely for people but for training the complex information systems too. Machine Learning systems and Artificial Intelligence software need to be trained, calibrated, and validated with data. From identifying pictures correctly to autonomous driving and labeling videos, and transcribing speech, people need to train the intelligent software, which has a lot of learning to do, before it can be used in real life. As software become progressively more intelligent, even these basic human interventions will be automated, and will be needed only for fewer but more sophisticated situations.

The skills, knowledge and experience of people, teams *and systems* should be constantly assessed and tracked, for both, using them better in the organization, and for their growth and development. This must be tracked centrally and regularly. People are the ultimate resources of an organization, more than financial or virtual/physical assets.

(b) *Continuous Upgrade of Systems and Processes*—All existing systems, tools, and frameworks need to have a roadmap for upgrade, or there is a real risk of becoming obsolete and uncompetitive fast. Integration and compatibility are of great importance. Upgrading one component should not lead to mismatches and failures elsewhere. In the new age, the capability of the entire organization rests on digital systems and processes. *Hence the power of an organization is limited by its systems.* Systems that are falling out of step with the industry must be sunset (i.e. phased out) in a planned manner. All contracts with vendors must reflect these organizational policies on upgrades, replacements, and maintenance. These should be factored into the cost structure of the organization. Of special concern are the cases where the vendor's organization is being acquired and products being shelved or not well supported.

(c) *Continuous Modeling and Automation*—The organization must keep evolving its models at all levels—from strategic to operational. With innovation arises the need to build new models. The organization must keep experimenting with new models on a smaller scale, before adopting them. Thus testing and continuous improvement should be ingrained in the organization's culture. When a model has to be adopted, it should be ready from the point of view of an information model that can be automated to the greatest extent possible.

(d) *Continuous Research and Innovation*—Both open innovation and closed innovation need a research base, even to scan properly the existing entities in the industry and to check for suitability for collaboration or adoption. Research should be an inherent part of the process even at the level of operations. For example, operations data can be analyzed every month for patterns, anomalies, and insights. The research on competitors, customers, and partners is of paramount importance. Beyond that, it is good to study players outside one's zone, for example, those in the same field but geographically distant, or those in parallel fields. A distant, uncaught trend may be adopted silently by a competitor. It's crucial to note that research must cut across the rank and file of the organization, to be an inherent part of every procedure.

(e) *Continuous Collaborations*—An organization that is closed is dying surely. It is good to adopt the Quadruple Helix Model that is explained elsewhere in the book. All partnerships and collaborations must be monitored and tended to carefully. And the organization must be on constant lookout for new kinds of collaborations.

(f) *Continuous Marketing*—Unlike in the past era, it is now possible to do focused marketing 24/7 across the globe, without much manpower or expertise, by using digital marketing tools, such as intelligent bots, automated chat assistants, web-marketing tools such as those that display different content depending on the geo location (IP address) of the person browsing the web page, and location-based mobile services. All these can gain meaningful information and insights about potential customers to facilitated targeted marketing and sales. Deal making and sales happen round the clock and large parts of it should be automated. Artificial Intelligence helps in automated matching of needs in the market and in putting up a pipeline of well-curated proposals that only need minimal approval in a workflow fashion.

(g) *Continuous Deliveries*—The new organization makes deliveries of products and services round the clock, and keeps refining them through Machine Learning, automated configuration, customization, and error correction. It takes into account the changing needs and varying tastes of customers, right down to an individual-customer level, that is, highly personalized. It does this by tracking a lot of data that it gets from a variety of sources, and from the individual profiles.

(h) *Continuous Recruitment*—This will be very different in the new world, where regular jobs are few and most are contract "gigs" (consulting assignments that are fee-based), or jobs outsourced to vendors. The manpower needs and the skillsets required will be extremely dynamic. Hence this must be undergo near-complete automation. Recruitment will happen round the clock, and very often there will be a "supply chain" management, that is, managing the list of projects/tasks to be delivered vs. the supply of manpower available as per the schedule/needs of the project. There will need to be adequate buffers. Recruitment will be a complex mix of hiring people on contract/permanent basis and outsourcing of jobs to other organizations. The latter is like "recruiting" an organization. The schedules, contracts, finances, deliveries, and skillsets will need to be managed all the time. They need to continuously align with the goals and KRAs, even through all the changes demanded by customers or necessitated by competition, and also account for the inevitable problems/slipups.

3.6   The Nature of Work

Knowledge bases, automation, analytics, Machine Learning, Artificial Intelligence, robotics, a distributed workforce, specialization all mean that the human quotient of work will progressively reduce to the most minimum possible. But it will also mean that work itself will be more clearly defined and measured. There will be deep study and research in every field to build models ranging from simple to sophisticated, to determine:

- *The Work Content*, that is, what a piece of work in that area really means, WBS (work breakdown structure), what it consists of, the parts, interrelationship between the parts, the outcomes, the static and dynamic behavior of the work output, the quantum of effort, and time needed for the work

- *The Value in Every Piece of Work*, as measured in monetary terms, its place in the value chain, and in relation to product/service roadmaps, contractual requirements, regulatory compliance, service/customer deliveries, stakeholder satisfaction, and organizational metrics
- *The Metrics* to track progress in work, which include metrics for project management, quality, financial, and so on

Entire categories of work will get eliminated rapidly in the years to come, owing to ICT and innovation. Technology is making redundant even advanced consulting roles. New categories will arise but more slowly. It is nearly impossible to predict exactly as to what could be the future growth areas, the precise nature of work, technology, business, or the tools. A couple of decades ago, the best brains in the world could not predict the full ramifications of the Internet, such as entire businesses, like eBay/Flipkart/Monster going online on a global and industrial scale. However the following fields seem to hold promise (Table 3.1).

Intelligent Technology will simply eliminate jobs that human beings are ill-suited for, such as repetitive jobs that need high precision and consistency. With increasing AI/ML, even messy laborious jobs can be automated. On the flip side, there will be a large workforce of human trainers needed to teach software through samples, use-cases, and correct examples, in every single area wherever complex software and AI/ML are used. As the latter will be all pervasive, so will be the need for human intervention.

Joseph Pistrui, a professor at IE Business School in Madrid, writes in *Harvard Business Review*, that great many jobs will be lost (even in the knowledge sector), and finally concludes that the future of human work is imagination, creativity, and strategy (Pistrui 2018).

However, as software becomes more intelligent, the nature of the human intervention will also shift from basic to advanced, and for only fewer complex scenarios. For example, Google created a neural networks software that could automatically identify cats (Markoff 2012). The software was shown thousands of samples of different cat videos, without explicitly tagging them as cats. And finally when a completely new cat video was shown, it identified it correctly. No training by humans was needed. Essentially, the software taught itself how to recognize cats. Google needs this because of the millions of YouTube videos it needs to classify on a daily basis. Fast forward to 2017, and now Google is trying to classify 5000 species of plants and animals (iNaturalist 2017; Gershgorn 2017).

Table 3.1   Promising fields for careers

S. No.	Field	Details
1.	Computing, Engineering, and Information Sciences	Computer science, ICT, data sciences, analytics, Artificial Intelligence, Machine Learning, robotics, IoT, natural language processing, automation, cloud and mobile computing, microservices, containers, social media, software engineering, enterprise architecture, technology architecture, design thinking, product management, solution design
2.	Cybersecurity	Security of websites, portals, transactions, services, identities, IT infrastructure, networks, public utilities and infrastructure (such as water supplies, sewerage, electrical grids, traffic control, dams, trains, airports, hospitals, etc.), industrial infrastructure, buildings, structures, banks, financial institutions, defense, government bodies, and so on
3.	AR/VR/Geo	Virtual Reality, augmented reality, geo informatics, geo spatial engineering, multimedia
4.	Media, Entertainment, and Gaming	Gaming for entertainment, education, training, tutoring, testing and recruitment, for model building and verification, serious games and applied games used in industries, for example, flight and medical simulation, multimedia and entertainment
5.	Life Sciences	Bio-engineering, biochemistry, biomedical engineering
6.	Health	Health technology, alternative medicine, and therapy
7.	Agriculture and Food	Urban farming, sustainable agricultural and food technology, ultra-efficient farming, greenhouse farming
8.	Financial Technology	Cryptocurrency, international finance, e-commerce, m-commerce, online banking, microservices financial transactions, virtual banks
9.	Transportation	Smart vehicles, autonomous driving, drones, logistics, vehicles powered by electricity, compressed gas, and hydrogen
10.	Energy	Alternative energy—solar, hydrogen, wind, tide
11.	Nano Technology	Graphene, micro-devices, carbon nanotubes, micro-robotics
12.	Sustainability	Environment protection, pollution control, water conservation, social responsibility, economic practice, sustainability innovation, next-generation infrastructure, climate change, and global warming
13.	Smart Cities, Smart Villages	Smart connectivity, smart public utilities, waste management, housing, smart buildings, management of space, energy, traffic, and so on, governance, law and order, maintenance, events management

(continued)

Table 3.1   (continued)

S. No.	Field	Details
14.	Legal	Intellectual property, contracts, international law, cyber regulations, risk, and compliance
15.	Digital Marketing	Marketing through all means on digital media—email, web, mobile. Affiliate marketing, social media, search, display, video, visual, and so on
16.	Aerospace Engineering	Avionics, drones, aero structures, astronautics, and so on
17.	Defense and Police	All areas, including those dealing with cyber terrorism, cyber attacks, distributed intelligence, image and video analytics, social media monitoring, and so on
18.	Governance	E-governance, innovations like crowdsourcing for citizen security, analytics and data mining for optimizing resources, tracking corruption, fraud and tax evasion.
20.	Educational Technology	MOOCs, web classrooms, personalized instruction over the net, intelligent tools for automated training, guidance and correction, self-assessment, tests, for example, for teaching programming. Training Artificial Intelligence systems, not merely training people. Validation and verification of models will be crucial
22.	Fundamental Sciences and Engineering	Specific areas of fundamental sciences will have good potential in the industry in the coming decades. Too many to list here: materials science, communication theory, molecular biology, econometrics, social science, cryptography, quantum computing, and so on

Source: Author's creation

The artistic fields are also influenced heavily by technology. While machines can tirelessly draw hard-coded artistic patterns with consistency, new tools like GNN (Generative Neural Networks) even create completely new patterns on an infinitely recurring basis. An older analogy of this is the kaleidoscope. However, the GNN type of tools are far more advanced.

The following human faculties will continue to be highly regarded and will dominate the work, more deeply than ever before, and with the help of tools of an ever-increasing sophistication (*The Economist* 2015). A Google-sponsored study by *The Economist*'s intelligence unit listed the following as the top skills for the future:

1. *Problem Solving*—analytical skills, mathematical and logical skills.
2. *Team Working*—integration, synthesis, holistic thinking, systems thinking, synergy.

3. *Communication*—effective oral and written communication skills, languages, foreign language skills.
4. *Critical Thinking*—The objective analysis and evaluation of an issue in order to form a judgment (Oxford Dictionaries). Abstract thinking, study, and research deep dive into the concepts, simplifying complexity, rethinking the fundamentals in different ways (Economist 2015).
5. *Creativity*—innovative thinking, seeing unobvious patterns, finding unconventional solutions, artistic talents.
6. *Leadership*—personal development, relationship building, social skills, team building, leveraging diversity, partnerships, and collaborations.
7. *Literacy and Digital Literacy*—technical/ICT/computational skills and expertise, dexterity with tools, especially software, programming, ability to learn new skills rapidly, mastering the learning curve.
8. *Emotional Intelligence*—mental and physical toughness, perseverance, strong will, confidence, breaking through tough barriers, spiritual strength, emotional intelligence, empathy, mental balance, humility.
9. *Entrepreneurship*—business sense, sports, gaming, working past failures, opportunism.
10. To add to the above list, ultimately, the key drivers for success in the long run are: *Domain/Industry Knowledge*—vast and deep knowledge in a field, interrelationship with other fields, cross-domain expertise, T-shaped knowledge—wide across several areas and deep in one area, financial literacy, management, and economics, especially for one's area and profession.

NASSCOM in partnership with BCG has identified specific future job roles in ICT and the related skillsets, in order to prepare India for future growth (NASSCOM Future Skills 2018). Organizations should take note and invest in their employees, training for the future skills of interest.

Tools will continue to magnify the human faculties tremendously and compete with them, to the extent that the tools will quickly outpace and replace the human roles in practically every area. Hence every person will need to build his education and hone his skills keenly in all the eight areas mentioned above, right from childhood, *and learning throughout life*. Unlike in the past, the chances are extremely high that *all* the areas will be demanded of *every* person in varying degrees over time. Focusing on only some of these areas is a strategy that used to work in the past, but to thrive and be *really* successful in the new era requires *all ten* of them, in different proportions depending on the specific field. Otherwise technology can quite easily replace the human role. Even the artistic fields will depend a lot on technology. Hence it is essential that such an integrated/holistic development commence right from childhood through schooling, adulthood, continuing throughout life. There will be no artificial demarcations such as learning years, earning years, and "yearning" years (retirement period).

3.7   Case in Point: Innovation in ICT with Digital Marketing

With the increased convergence of ICT with business management fields, marketing is perhaps the most dynamic in terms of adopting new IT and communication technologies. This is helped by the fact that digital marketing today is at the forefront of handling customer touchpoints with brands. Customers first go online and look up companies, products, and services, and digital (along with word of mouth) has accelerated the awareness of customers with every possible choice available.

3.7.1   The Rise of MarTech

To keep pace with rapidly changing customer tastes, habits, and digital behavior, customers are increasingly adopting MarTech (marketing technology). Sensing the opportunity, there are more than **5000** vendors in this space offering a variety of products from marketing automation platforms to niche MarTech tools.

MarTech platforms like Qwardo (http://qwardo.com) provide innovative, AI-powered content engagement base to deliver personalized buyer journeys to increase high-quality leads and returns on content. They provide intelligent engagement tools that use an underlying analytical foundation to deliver content recommendations and call to actions. Such engagement techniques improve the engagement and lead generation by maximizing the content utilization and providing the data-driven approaches to content marketing.

Their overarching goal is to empower marketers with data and analysis of all the user behavior that is now being captured across channels and deliver hyper-personalized user experiences to drive brand loyalty and revenue, as well as feed product development cycles. The modern tools are cost-effective platforms that focus on getting the most out of content marketing and delivering sales-ready leads.

3.7.2   Applications in Marketing

As with other industries, marketing technology is also being disrupted by AI and advanced data analytics. So far, the most commonly adopted innovations have focused on automating marketing activities, but not so much on customer experiences with brands.

With AI, marketers can increase the engagement of content and calls to action that drive customers to learn more about brands and services they care for and convert as high quality of leads. The MarTech platform leverages Artificial Intelligence algorithms to deliver relevant content recommendations to website visitors based on their persona and where they are at in their journey with the brand.

3.7.3   Trends in B2C Marketing

Marketers in the B2C world lead the trends in adopting AI and related technologies as the amount of data they gather is immense and end-users of B2C products and services are very demanding. Besides the likes of Amazon and Netflix, mid-sized companies can benefit from personalization platforms such as Qwardo by engaging end-users on their site based on user behavior, consideration, and other parameters with dynamic offers and recommendations. Shoppers and other kinds of consumers have plenty of choices online to hop from one brand to another and have scarce attention spans. Hence the need for dynamic and personalized engagement.

Delivering the right offer to each consumer at the right time and place is key. With one-on-one marketing capabilities, there is an infinite set of ways to personalize offers for each user based on purchase intent. You can create discrete segments and deliver personalize offers or let the machine determine the best product recommendation based on data it gathers for each user. Most of the customer support interactions will be handled by intelligent chatbots.

3.7.4   Trends in B2B Marketing

In the B2B world, purchasing cycles are complex and long. Decision makers do a lot of research before talking to the sales team of a vendor. Marketers align with this process by offering educational content such as research reports, white papers and case studies on their sites to aid decision makers.

B2B marketers deal with discrete customer segments and create content for each segment and each step of the buyer's journey. Here the key is to ensure delivery of the right content to each segment at the right point in their journey, and accelerate sales cycles, while delivering account intelligence and qualified leads.

The MarTech platform analyzes content consumption at the engagement level and can determine which content is working well for which segments, and deliver the most engaging content to each prospect. You can configure poll questions to get quick feedback and segment users. Signup CTAs (calls to action) can be used to increase conversions across the site, and onsite retargeting can dramatically increase user engagement rates by delivering targeted messages to campaign leads.

3.7.5   Content Marketing: A New Paradigm

Using strategically positioned content to drive business has now become a reality, with most marketers investing heavily into research, content creation, and distribution. Blogs are regularly published; newsletters and social media channels are populated with white papers. However, the quality of lead generation has always been a sore spot for content marketers.

3.7.6   Proactive vs. Reactive Marketing

With innovations in campaigns, search engine optimization techniques, and the explosion of social media, driving traffic to content heavy sites has been mastered. However, once the visitor is there on the site, not much can be done to influence the buyer and capture feedback. Here the opportunity is huge, as buyers consume relevant content from a brand, it increases their brand loyalty and leads to a purchase.

Rather than waiting for signals from the buyer (lead forms, contact us forms), marketers can proactively gauge the interests of buyers and deliver the right content before losing the prospect from their sites. An always-on platform like Qwardo can make this process seamless and user friendly, without requiring a lot of time and effort from Marketers.

3.7.7 *Personalized Engagement at the Segment, Company, and Contact Level*

With an intelligent B2B marketing platform, personalization can be delivered at various levels of segmentation. While earlier you could offer dynamic experiences by location, today you can group visitors by industry, account names, campaigns, user behavior, and interests. Once the visitor is on the website, classification is done automatically, and a personalized content recommendation can be delivered to the visitor.

This keeps the buyer engaged and reduces bounce rates where one loses the visitor from the very first page that they land on. The entire user engagement can be completely automated, with content recommendation engines using user behavior to deliver relevant content. Or, marketers could design a custom experience for each customer segment.

AI-driven content engines increase the accuracy of segmentation and in delivering content that is working for different segments.

3.7.8 *Measuring Content Performance*

As more and more content is published, the shelf life of content is reduced. Marketers focus on promoting newer content and older content is left underutilized. Before creating new content, marketers could instead focus on understanding what kind of content is working well and with whom. This analysis goes beyond measuring clicks and downloads, to understanding the engagement of qualified prospects with content, that is, completion rates and time spent in consumption.

3.7.9 *Costs*

Marketers run campaigns, create, and manage content, and, among all such related activities, they have to deal with dependencies on IT and budgetary concerns for new technologies that are initially built for the enterprise market and priced out of the reach of SMEs.

3.7.10   Reduced Dependence on IT

Implementing MarTech can be complex, and integrating with other parts of the existing IT and marketing systems needs close support from IT and security teams. Once approved, many marketing technologies require IT support on an ongoing basis to make changes.

MarTech tools address this problem by offering a simplified paradigm that allows marketers to take advantage of the product completely with minimal dependence on IT. They also connect to state-of-the-art web-based workflow automation systems such as Zapier (https://zapier.com) to enable data exchange between the MarTech tool and other leading marketing automation and CRM vendors.

3.7.11   The Bottom Line

Marketers need more tools to deliver predictable and measurable business outcomes. These tools need to be intelligent and shouldn't require a lot of expertise to use them. Laggards in adopting and learning new technologies will only fall back in engaging and retaining customers while the rest of the industry goes through trials and tribulations in learning how to get the best of results out of their investments in MarTech.

Marketing Technology tools like Qwardo and HubSpot (www.hubspot.com) offer an intelligent content engagement platform that can help marketers drive more utilization of their content and thus increase return on their investments while delivering an engaging and personalized content experience to the website visitors.

3.8   Conclusion

In the Information Age, information should be treated as the key asset of an organization and should occupy the top of the value hierarchy. Failure to truly recognize this fact will render a company uncompetitive and will rapidly drive an organization into obsolescence and closure. It is important to note that the value chain, the nature of work, and the kind of skills that are valued are changing fast. Digitalization or digital transformation is therefore the most important undertaking in an organization. This transformation needs to be driven by the focus of an organization, the information valuation, innovation strategy, the capability, collaborative potential, and social capital required of workforce. Business models, intellectual property, development and operations processes, technology, innovation are all no longer static entities but are fast evolving dynamic entities. The frameworks and platforms built for an organization should enable and spur these capabilities.

Information or data processing can be done on at least three levels depending on the company's priority: at the operational level, tactical level, or strategic level. At the operational level, basic information are processed at the day-to-day level for process efficiency. At the tactical level, summarized data helps middle-level managers to take decision at the process level such as any tweaking or mid-way correction to achieve certain targets. Data can be used for strategic decision at the top management level such as product innovation based on customer need or new services to fill up a gap. In the presence of rapid technological change, information processing at these three levels can be linked for a continuous mapping which can give direction to a company to remain ahead of others to beat the competitive pressure.

The Takeaway Box: Information Modeling, Digitalization, and Org Development Are the Keys to Execution

- Look at everything through the lens of information sciences, digitalization, and automation.
- Think of every organization as an information processing engine.
- Nurture and develop the organization and individual talents to prepare for the new era.
- To build a resilient and successful organization, invest in the skills of the future.

REFERENCES

Birudavolu, S. (2015). Open Innovation in ICT: An Empirical Assessment of Global Telecommunications Services. Unpublished thesis submitted for PhD at Indian Institute of Foreign Trade, New Delhi.

Economist. (2015). *Driving the Skills Agenda, Preparing Students for the Future.* An Economist Intelligence Unit Report sponsored by Google. Retrieved from https://static.googleusercontent.com/media/edu.google.com/en//pdfs/skills-of-the-future-report.pdf.

Gershgorn, D. (2017, April 11). Five Years Ago, AI Was Struggling to Identify Cats. Now It's Trying to Tackle 5000 Species. *Quartz.* Retrieved from https://qz.com/954530/five-years-ago-ai-was-struggling-to-identify-cats-now-its-trying-to-tackle-5000-species/.

iNaturalist Challenge. (2017). *The Fourth Workshop on Fine-Grained Visual Categorization*. Retrieved from https://sites.google.com/view/fgvc4/competitions/inaturalist.

Kay, John. (1995) Foundations of corporate success: how business strategies add value. Oxford Paperbacks.

Laney, D. (2017, November 13). Turn Your Big Data into a Valued Corporate Asset, Gartner, *Forbes*. Retrieved from https://www.forbes.com/sites/gartnergroup/2017/11/13/turn-your-big-data-into-a-valued-corporate-asset.

Markoff, J. (2012, June 25). How Many Computers to Identify a Cat? 16,000. *New York Times*. Retrieved from http://www.nytimes.com/2012/06/26/technology/in-a-big-network-of-computers-evidence-of-machine-learning.html.

NASSCOM Future Skills. (2018). Retrieved from http://futureskills.nasscom.in

Oxford Dictionaries, Definition of Critical Thinking. Retrieved from https://en.oxforddictionaries.com/definition/critical_thinking.

Pistrui, J. (2018, January 18). The Future of Human Work Is Imagination, Creativity, and Strategy. *Harvard Business Review*. Retrieved from https://hbr.org/2018/01/the-future-of-human-work-is-imagination-creativity-and-strategy.

Senge, P. M. (1990). *The Learning Organization*. Doubleday/Currency, ISBN 0-385-26094-6.

Viviano, F. (2016, September). How This Tiny Country Feeds the World. *National Geographic Magazine*. Retrieved from www.nationalgeographic.com/magazine/2017/09/holland-agriculture-sustainable-farming/.

CHAPTER 4

The Enterprise Technology Landscape

Technology is a useful servant but a dangerous master.
—*Christian Lange, Nobel Peace Prize Lecture, 1921*

4.1 INTRODUCTION

Conceiving a business model that has an excellent fit with the market has no value unless it is executed well. One should be familiar with the enterprise technology landscape before attempting to implement any innovation or a new business model; otherwise there is a risk of failure or of being run over by the others who have been smart enough to understand and leverage the new means effectively. There is, in fact, a twin-fold danger. One is that of losing the edge to the competition, and the other that of sinking scarce resources into something unnecessary or obsolete. Sometimes overambition takes over and one embarks down the path of adopting a cool, new, cutting-edge technology that is the buzz of the town. But this soon becomes the nemesis of the business, as the new technology is beset with unforeseen problems. And the business becomes a guinea pig for the unproven new technology.

A proven new technology could have accomplished the same (and much more) with a great deal of simplicity. Once a path has been taken without studying the best-of-breed technologies, and investments deployed, it is hard to retract from it easily because commitments have been made on every side. It also becomes progressively harder to change

© The Author(s) 2019
S. Birudavolu, B. Nag, *Business Innovation and ICT Strategies*,
https://Doi.org/10.1007/978-981-13-1675-3_4

63

Table 4.1 Applications and technology for IoT/M2M based on geographical spread vs. mobility

		Fixed	Mobile
↑	Dispersed	**Applications**	**Applications**
		Smart city/campus	Car automation, e-Health
		Smart metering, smart grid, remote monitoring	Logistics, portable electronics
			Fleet management, Field workforce mgt
Geographical spread		**Technology**	**Technology**
		PSTN, broadband, 2G/3G/4G/5G	2G/3G/4G/5G
		Fixed wireless	Mobile broadband, SMS
		Powerline communication	Satellite Communications
	Concentrated	**Applications**	**Applications**
		Smart home/building/colony	Onsite logistics, VIP security
		Factory automation	Visitors' tracking in a building/campus
		Hospital equipment management	Event management
		Technology	**Technology**
		Wired personal area network	PSTN, broadband, 2G/3G/4G/5G
		Wired network	Mobile broadband, SMS
		Indoor electrical wiring, Wi-Fi	Powerline communication
		Fixed wireless	

Need for mobility →

Source: DoT, National Telecom M2M Roadmap, May 2015

direction and move towards the new technologies. A double whammy indeed!

On the flip side, a company waits for too long, struggling to work with its legacy systems that are increasingly difficult to maintain, upgrade, and integrate, leading to failure of the business at multiple points. The organization should have embarked on a transformation project years ago, adopting new but proven technologies. To give an example of the way the landscape of technologies evolves, Table 4.1 lists the IOT/M2M applications and technologies along the dimensions of geographical spread versus mobility.

Of the hundreds of emerging and established technologies, which ones will gain strength? And what are the cautions? This chapter helps in culling out the key technologies that will play a major role in transforming the industry for the coming years (and decades). Many of these may seem known and widely spoken about, but it is surprising how many people (and organizations) in the industry still do not realize their power and are still hesitant to embrace them. It is hoped that this chapter will help move their center of gravity away from obsolescence towards a bright promising era of agility, growth, flexibility, and innovation, which the new technologies help leverage in abundance. One may have to restructure or write off old investments to embark on this path, and that is OK, considering the long-term vision and immense gains. The effort, the short-term loss, and pain will be compensated many times over if one plays the cards right.

4.2   THE BROAD THEMES

The coming decades will bring revolutionary changes in every sphere, with far-reaching breakthroughs in every field, and driven by the magnifying/analyzing/combining power of ICT (Birudavolu 2015). Let's look at the broad themes and explore why they've gathered strength.

4.2.1   Cloud Computing

This section should really be unnecessary, but for the fact that it is shocking to see that many in the industry still clinging to on-premises IT infrastructure and computing, for whatever reasons (sunk costs, biases, ongoing status quo practice, known expertise, existing teams, political environment, existing contracts with IT vendors, security reasons, etc.).

ICT has gotten immensely complex and very large. The configuration, operation, and management of the IT infra is an *expert's game*. ICT is much too specialized and expensive to expect any organization to take on, on their own, unless, of course, ICT is the organization's mainstay. The industry currently demands agility and flexibility, which means that business models change frequently. Consequently, the business applications and IT deployments will need to keep pace with the changes and retain the ability to execute the strategy and operations of the changing business.

All these reasons indicate that it is far better for normal organizations to pool their ICT resources and entrust specialized ICT organizations with managing and operating these ICT resources on their behalf. With adequate policies and processes in place, the client organizations should be able to safely access their business applications and data and focus their energy and resources on delivering their main business value proposition.

In fact, the IT game is very hard even for ICT organizations themselves. Hence, they invest heavily in automation for dealing with the complexity and scale, to be able to service the SLAs of their client organizations. If you have any doubts left, consider the following.

Here is a quick recap of the layers (loosely classified):

- *Services Layer*: business operations, customer journeys, self-service portals, customer support centers
- *The Application Software*: middleware and databases, business applications accessible through browsers, client software on PCs, mobile devices
- *The Network*: switches, routers, Wi-Fi hubs, LAN (local area network), MAN (metropolitan area network), WAN (wide area network), SBCs (session border controllers), VPNs (virtual private networks), SDN (software-defined networking), NFV (network function virtualization), EPABX, VoIP server
- *The IT Software*: the operating systems, the virtual layers, tools
- *The Hardware*: computing servers, storage, PCs, mobile devices, IoT, gaming hardware, televisions/display screens, AR/VR devices, surveillance cameras

As if the complexity of the layers were not enough, the demands and expectations are growing by the day:

24/7 availability (5 Nines, or 99.999% uptime), planet-scale deployments, shrinking lifecycle of products and deliveries, seamless integration in development and operations processes, distributed deployments across the region or globe, high security of data and transactions, flexibility to cater to growing needs and changing business models.

Cloud is an umbrella term for all shared pools of ICT computing resources that can be accessed remotely 24/7. It is considered a utility, in the sense of metered service, pay as per usage, scale as needed, packaged/ bundled/tiered services such as compute + storage + memory, and so on. The cloud offerings are service oriented, such as:

(IaaS) Infrastructure as a Service, SaaS (Software as a Service), PaaS (Process as a Service), KaaS (Knowledge as a Service), and so on.

The enterprise customers of the cloud typically set SLAs for security, response time, availability, scalability, support, speed of deployment processes, and so on. For further reading, see (cloud). Popular cloud services in the industry are Amazon EC2, Google Cloud, Microsoft Azure, and Oracle Cloud. Each of these cloud services provides an entire environment replete with DevOps tools, storage space, servers, geo-redundancy, choice of myriad operating systems, inbuilt security, fault-tolerance, web-hosting utilities and services, inbuilt metrics and analytics, dashboards for ease of regulatory compliance, and so on.

In the layers described, it is best for organizations to largely outsource off the layers, as follows:

- Hardware, IT Software, and Network—To the cloud service providers and cloud infrastructure management companies
- Application Software—To the system integrators, who will develop and deploy solutions
- Services—To the managed service providers, who run business services and operations on the cloud on behalf of the organization

Typically, some large system integrators can take on full responsibility in a blanket contract, and they manage the other vendors on behalf of the customer.

Organizations can thus maintain a fledgling in-house staff for ICT infrastructure management, and remain focused on the higher functions, such as business strategy/management, information sciences, and so on.

Cloud is the clear choice for computing and should be the default mode of operation in every organization, exceptions notwithstanding. Unless there are compelling reasons, one should never opt for on-premises servers, as it brings with it a load of responsibility which the organization is likely to be in no position to handle. Organizations with existing on-premises solutions should immediately plan in their roadmap a strategy to migrate to cloud. Hybrid solutions, which are a mix of on-premises and cloud, also don't work out so well in the long term as there is the added headache of dealing with complicated architectures, operations, implementation and upgrades, and dangers of security loopholes. They should be treated as an interim solution at best. It is best to cut down on-premises solutions to zero or if there are compelling reasons restrict it to below 5% of the entire ICT deployment. A good example of a real need to keep an on-premises device is a Cisco edge router that aggregates IoT data and does on-premises computation to manage the IoT sensors locally and reduce Internet traffic, and also for providing cybersecurity to vulnerable devices by standing between them and the Internet. It has the intelligence to ward off malicious intrusions from the Internet.

Thus, it is essential to build cloud native applications, rather than building for on-premises and retrofitting to cloud. This will provide superior speed and resilience in the applications. For best results, applications should be built cloud native, with a distributed architecture, and a design based on microservices, with the processes being integrated DevOps. This will be the norm going ahead.

4.2.2   Edge Computing

Edge computing is a way to simplify the Internet by doing part of the computation on the edge of the Internet instead of doing everything on the central servers, that is, the nodes. The edge is defined as the border of the Internet which makes contact with the physical world, as against the central servers that are deep inside the Internet and have no direct contact with the world, that is, they're only connected to other devices. Examples of edge computing include wireless sensor networks, grid/mesh computing, mobile edge computing, AR/VR, distributed data storage, and IoT.

Even with low-cost, high-speed broadband, sending information over the Internet is an overhead. This is felt more acutely when the Internet will fully encompass IoT also, thereby becoming IoE (Internet of Everything). Millions of sensors and devices will be connected to nodes (servers) over the Internet. In many cases, the data is sent to the backend for processing so that the output of the processing (e.g. a decision or computed result) can be sent back to the devices for usage by the end-clients or appliances. For example, in surveillance through video cameras, the images may be sent to the backend so that the server can match them against some images from database and against some rules. With inexpensive computing power and storage, most of this processing can be done locally at the site of the video cameras, rather than sending them to a remote server. This would save a great deal of Internet traffic. The video cameras are hooked up to the local edge computing device, which caters to the bulk of the surveillance needs, reaching out to the remote central server only when needed. The edge computing device maintains local storage and computing and keeps updating its local database and business rules occasionally on a need basis.

Many companies like Intel and Cisco have developed intelligent edge computing devices.

4.2.3   Mobile Computing

Mobility is also very obvious and is widely known and used and hence can't really be called a future technology. However, it merits a small note at least. Mobile computing offers an unimaginable degree of portability, connectivity, personalization, and social interactivity. The rich array of mobile devices spans the range from wearable computers to laptops, smartphones, and tablets. The ecosystem of mobile applications, incentivized through open innovation, caters to practically every need and demand of the individual and the business. Mobile computing should form an integral component of any enterprise architecture. Enterprise mobility provides mobile computing solutions for the enterprises, such as automation for the sales personnel in the field. For example, they may be able to access their CRM or sales deals through their mobile device. Many software applications are mobile-only or mobile-first, focusing on access through mobile phones rather than through desktop PCs, because the market penetration of mobile phones is far higher than that of PCs. LBS (location-based services), mobile money, social media, personal productivity applications, health applications are but a handful of the variety of applications that have revolutionized the use of ICT in societies across the globe.

The 5th Generation Wireless System, abbreviated to 5G, offers much faster connectivity, up to 20 gigabits per second and, with very low latency, will greatly benefit the bandwidth-thirsty IoT, automotive, and industrial sectors. It will certainly play a big part in commoditizing bandwidth.

However, 5G will need massive upgrades in infrastructure, including adding more cell sites and equipment, which all mean denser deployment. The first deployment of 5G is scheduled for 2020.

4.2.4   Platform Architecture

Platforms will drive ecosystems. They are both a current and established technology as well as an evolving and futuristic technology that will progressively get more intelligent. Platforms have been described in a more detail elsewhere in another chapter.

4.2.5   Microservices and Containers

Microservices is all about slicing up a cloud software application into fine-grained units to get business features to the market rapidly. Each microservice is a bare minimum (smallest) unit of functionality that can be deployed separately but, when combined with other microservices, can offer a larger business use-case functionality. The flexibility and scale required by the current-day business models demand an entirely new kind of software design that is based on SOA (service-oriented architecture). Microservices is a form of SOA. The benefits of microservices are faster and more robust deliveries, lower cost, easy to distribute an application (i.e. do distributed computing), easy to offer scalability and high availability, makes service orchestration simpler, testing and administration can (and should) be fully automated, service registry, and service discovery.

The key characteristics of microservices software design are:

1. Modular design, fine-grained components, each delivering a service, for example, inventory microservice, payment microservice, product catalog microservice, and so on. Each microservice component has the following layers:
   (a) *At the top layer*: An API exposed by the microservice
   (b) The application that handles all the API's requests
   (c) A data store
   (d) *Bottom-most layer*: An infrastructure component
2. Loosely coupled components that are easy to mix and match to deliver different kinds of services.
3. Lightweight protocols that enable components to be combined and orchestrated to deliver a service.
4. Integrated DevOps, enabling continuous delivery of services, each independently deployable, minimize integration effort and testing, incremental changes should require minimal changes/testing restricted to a small number of components.
5. Business-/domain-driven design and development—The primary focus of the entire development project is on directly implementing the business model and business rules. The software design and implementation is a direct representation of the business domain/model's complexity.

6. Development team's organization—One development team per microservice. Agile development process. There are no central committees like in the development of monolithic software applications. The turnaround time of a microservices-based delivery is often orders of magnitude faster than the traditional software deliveries.

Examples of microservices frameworks are Spring Boot, Spark Java, Akka.

A container is another important paradigm borne of cloud computing ecosystems which addresses the problem of portability of an application (containers). An application developed, tested, and deployed in one environment *may not work* in another, due to differences in the operating system, configuration, library versions, subtle differences in infrastructure, and so on. A container packages an entire run-time environment within itself, that is, the application software, libraries, configuration files, binaries, and so on. So, this container when deployed on different environments is *guaranteed* to run correctly and produce the desired output. There is no further development and testing needed. This is akin to standard sockets for bulbs, so that different bulbs regardless of their manufacturer, wattage, shape, size, color, or blinking pattern are able to fit in perfectly into the socket and perform (glow) as per the bulb's own specifications. There is no need to "test the fit" or "re-size" the interface of either the socket or the bulb, for using the bulb.

In essence, a container is a kind of platform for applications which guarantees that an application bundled into it will always execute well wherever the container is deployed. This, of course, assumes that the application itself has been developed and tested well and is free of defects, in the first place. The container assures that there will be no run-time problems because of the environment. So, it guarantees portability of the application, without having to test and change the application or any of its dependencies (configuration/libraries/binaries) for another platform.

A good practice is that an application built using microservices be delivered in a container. Although this is not mandatory, this is really recommended because it offers closer integration, lesser wastage, ease of orchestration and administration in terms of starting and stopping the application, and the biggest advantage is that the application becomes extremely portable.

Examples of containers in the cloud industry are Docker, Kubernetes, Tectonic, OpenShift Container Platform.

4.2.6   IoT (Internet of Things)

This is low-hanging fruit in the industry and its usage is all set to spike sharply. It is estimated that the global market value of IoT will reach US$7.1 trillion by 2020 (Hsu and Lin 2016). Table 4.1 lists the roadmap for IOT/M2M technologies classified by geographical spread versus mobility.

IoT is a natural extension of the Internet from a human-to-human (e.g. email, social media) and a human-to-machine/machine-to-human (e.g. web surfing, e-commerce, banking, ERP on cloud) to now include a machine-to-machine interface also. The difference between plain M2M (machine-to-machine) and IoT is that M2M can happen without Internet also, through wired or mobile networks and through embedded systems. M2M is mostly point-to-point communication and is the plumbing, whereas the IoT layer is a universal enabler. IoT uses the IP networks to connect the devices, typically to the cloud or to an application through a middleware platform. IoT data is integrated with enterprise systems to improve business by using sensor data in big data analytics for business, for troubleshooting, field workforce deployment, materials management, project management, surveillance, and so on. M2M is very much needed especially where there is no need to burden the Internet with a flood of sensor signals, which can be aggregated off or transmitted to another device/machine/server directly. Internet can then be used to transmit the gist of the data, or the aggregated data, or merely the exceptions. A lot of rule-based processing can happen before putting any data on the Internet. See the section on the Internet, to understand why it is not necessarily the best solution for all circumstances and at all levels.

IoT brings together the following four components:

1. Embedded electronics, actuators, sensors, for example, temperature, pressure, light, electric voltage/current/resistance that are embedded inside devices.
2. Devices, machines, for example, industrial machinery, field machinery such as construction equipment, mobile devices, PCs, vehicles, embedded systems, drones, home appliances such as electric meters, refrigerators, washing machines, and so on. With IoT these connected devices can be monitored and/or control remotely over the Internet.

3. Network connectivity—IP networks (essential), LAN, mobile networks—all of which allow the devices to interconnect and operate on the Internet.
4. Computing resources, for example, servers on cloud/on-premises, middleware, enterprise software applications.

IoT involves three parts:

1. Data transmission *from* sensors or *to* the devices. All of these are identified entities on the network. For example, several sensors measuring soil moisture in a crop at regular intervals.
2. Collection and analysis of this data, done by separate systems. For example, aggregating the data from different sensors, collecting them based on the crop/part of crop, and calculating the overall state of the crop's moisture
3. Decisions and actions based on automated processing, such as big data analytics, rule-based engines. This may involve a step that goes back to Part 1. Or the decision may be purely for monitoring and logistics purposes. The data can be fed into a Machine Learning system.
   For example, if the moisture is high in a crop, the decision is to turn off the water pump, but if the moisture is below a threshold level and X number of hours have elapsed, then switch on the water pump. This will need data transmission to the device that controls the pump.

The advantages of IoT are:

- Direct control, automation, and improvement in efficiency, accuracy, and timeliness.
- Superior modeling in the higher layers, for example, by using operations research to optimize traffic lights in an entire region of a city rather than controlling individual junctions can cause better flow control of traffic because the traffic lights function in coordination with one another based on intensity of traffic.
- Economic benefits due to improved logistics and reduction of wastage, for example, using a feedback loop wherein sensors measure soil moisture in a crop and the data is used to switch on or switch off a water pump, lead to superior results. Another example is that of vehicles enabled with IoT for tracking their location, vehicles' fuel levels and critical health parameters pertaining to engine, battery, etc.

Some more examples of deploying IoT are:

Remote inspection and monitoring of facilities and projects, e.g. oil rigs, electric transmission towers, mobile phone towers, wind turbines, construction projects, etc. Precision Farming, Tracking Inventory, Material and Assets, Operational Intelligence in Real-time E.g. Oil pipelines, cars in parking lot, goods in warehouse, people entering stadium, construction equipment in the field, raw materials being processed in factory, vehicle fuel monitoring, Customer Self Service, Performance benchmarking, Troubleshooting and Repair/Breakdown services, etc.

Industry standards for IoT are evolving. Eclipse Kura is an open-source framework that provides a platform for building IoT gateways (Kura). There are many protocols and frameworks built at different layers, for IoT, for example, see IoT Protocols.

4.2.7   Artificial Intelligence

AI (Artificial Intelligence) refers to human-like intelligence embedded within software. The world is approaching an age of super-intelligence (Urban 2015) with AI, which will augment and outrun human intelligence in most areas. AI is a general term that covers several areas, like awareness of environment, cognitive functions, learning, problem solving, decision-making, natural language processing, speech recognition, visual perception, and so on. The global market for AI is expected to hit $37 billion by 2025 (Statista AI 2016). And this may be a very conservative estimate, because the pace is picking up. Several countries are taking AI seriously. The Indian government has set up an expert panel on AI policy in December 2017. UAE has appointed a minister for Artificial Intelligence in October 2017. Russian President Vladimir Putin said on September 2, 2017, that the future belongs to AI and whoever masters it will rule the world (Meyer 2017). He also added that AI comes with colossal opportunities but also threats that are difficult to predict (RT 2017). AI is powerful and hence is a double-edged sword. It's necessary to have regulations and policies in place to prevent misuse.

In the World Economic Forum at Davos in 2018, Accenture released its Strategy Report in which it said that AI will boost revenues by 38% and employment by 10% by 2022, and it would boost profits by US$4.8 trillion globally in the same period (IANS 2018).

Organizations will perish if they do not consider AI in their strategy, while their competitors adopt it. As technologies evolve and adoption increases, and when the usage of a particular technology becomes routine, the capabilities of the technology are no longer considered intelligent, for example, grammar checking in word processing software. So, the term AI will always refer to cutting-edge technology. Stanford University led research team AI100 has launched an Artificial Intelligence Index (AI Index 2016) to track the state of Artificial Intelligence and measure technological progress, similar to how a stock index tracks the stock market. This is part of the One Hundred Year Study on Artificial Intelligence (AI100 2016).

AI is already gaining a large momentum and is poised to outpace normal human intelligence. This is both a threat and an opportunity, as AI will replace many jobs being done by people but will also create new ones. AI can now handle a wide range of jobs, which is increasing by the day:

From routine ones like support centers, rule-based troubleshooting, network planning, monitoring and optimization, more intelligent household devices (for lighting, heating, washing, cleaning, refrigerating, entertainment, health monitoring),

To difficult, dangerous, and dirty jobs—like cleaning up chemical spills, undersea/ocean floor exploration, crawling into remote, impossible to reach areas of underground mines/caves/structures/slush/marshes, hazardous areas like radioactive sites or air filled with poisonous fumes, firefighting, defusing explosives, clearing landmines, autonomous weapons, and so on,

To complex ones, like playing chess, autonomous driving, many consulting jobs, guided tutoring/training, investment decisions, risk assessment, cybersecurity, facial recognition, disease diagnosis,

And even entertainment and arts (Jones 2017). And in sports to manage players' performance and predict results of games and in planning strategy.

AI leverages several fields: Computer Science, Engineering, Mathematics, Linguistics, Psychology, Economics, Philosophy, Life Sciences, Medicine, Neuroscience, and so on.

AI is a vast and complex subject and has several branches, referred to as AI Branches, such as:

Symbolic AI, Machine Learning, Neural Networks, Fuzzy Systems, Probabilistic Methods, Evolutionary Computing, Soft Computing, Swarm Computing, etc.

It can be classified by approach or by application.

For example, Machine Learning is all about creating software that can learn on its own from the environment and the data it has access to. Machine Learning can use neural networks or genetic algorithms or statistical methods such as linear regression, clustering, classification, and so on.

A simple breakup of the vast field of AI could be based on its capabilities, ranging from the simple to the complex (Devbattles):

- Perception: Apart from the human range of perception, AI can have others like ultrasound, radar, x-ray, and so on.
- NLP (Natural Language Processing): Interpret and generate text and speech as needed.
- Knowledge Representation: Makes sense of entities and their relation, like objects, people, concepts, language, different subjects, and so on, and finds a way to represent them.
- Reasoning: Solve problems with the knowledge the system possesses.
- Planning and Navigation: To live in the real world, navigate its myriad complexities, and find paths/solutions.

As an example of large-scale unsupervised, deep learning, to mimic the human brain, Google built a neural network consisting of 16,000 processors and a billion connections and exposed it to 10 million different videos of cats (Google Cat Videos Recognition 2012). The system started learning on its own that these videos were showing cats. After the training, when the system was tested by showing it assorted videos, it could correctly identify which were cat videos, with a 75% accuracy, similar to human cognition levels. It's important to note that during training, the system was never ever told what a cat was nor was it given any video or photo tagged/labeled as cat. The entire training was run with unlabeled videos, and it was not told of any distinguishing features of cat. The system was entirely self-taught. When applied to human faces, the system could achieve 82% accuracy in recognizing them. Google could now use this to automate the process of classifying millions of videos uploaded into YouTube every day.

Different versions of this system can be used for rapidly identifying faces on surveillance videos in real time.

Another example is from Georgia Institute of Technology, where the AI Program was able to recreate a game like Mario just by watching it being played (Georgia AI 2017). It could deconstruct the entire game, extract the rules, and recreate the game engine, all on its own. Such advances pave the way for millions of real-life applications.

An AI application from Microsoft can generate images when given text (Ghoshal 2018), tested using GANs (generative adversarial networks). A GAN attempts to fool the AI application by spewing out fake data. This is to help push the limits of the capability of the application.

The company John Deere acquired Blue River Technology which uses deep learning in agriculture, to aid farmers monitor and manage crops and livestock. There is an explosion in the use of AI. For a few more examples, read the article (Huffington Post 2017)

AI can and should be used in practically every area. The premise is that every area of human endeavor can be improved to be made more intelligent than it currently is. A few examples are:

spreadsheets, gaming (e.g. chess, go), automated virtual assistants, fraud detection, strategic planning, decision support systems, project planning, customer support, traffic management, personal finances, wealth management, investment banking, healthcare, and so on.

New regulations are needed to account for the use of Artificial Intelligence (Microsoft AI 2018). There are many reasons, such as:

- When part or whole of the decision-making shifts to algorithms and machines, then who carries the responsibility for the consequences? For example, if an intelligent system causes an accident, due to defects, then who is responsible? The person in charge of the operations of the system, or the owner of the system, or the system manufacturer? Or is it the technology company that conceived the requirements and the algorithm? Or the party which quality tested and certified it?

    In practice, how will the liabilities be determined, and legal contracts drawn up? How will the courts resolve disputes and award penalties?
- When AI and Robotics replace human jobs, then will the tax burden shift to the systems? Or will it vanish?

- What about the compensation and benefits to the displaced workers? In the burgeoning gig economy, how will the work and compensation/benefits for the gig workers be quantified?
- How will regulations control the harmful use of AI? Such as those for intelligent autonomous weapons, breaching security, espionage, crime, manipulating public sentiments and opinion, rigging elections, devious and illegal means to seize markets and clobber competition, and so on. How will the limits be drawn, standards and practices be established, tools for monitoring them be built, and training be given to check compliance?

As is evident, AI will replace many human jobs (Upson 2016; Whitehouse 2016), but it will also create new ones (Wilson et al. 2017; Hutson 2017). The new jobs will be unlike any of the older jobs. Examples are:

- Training AI software, for example, for tagging for supervised learning, rating the output of the software to guide the software to learn correctly.
- Testing the software through thousands of scenarios to ensure that it works well in real life. This is more serious than testing ordinary software, because the AI software "learns" and "makes decisions".
- Running simulations and data generation, not only to give the AI system some real-life experience but also to trace back what the software has learned and improve it through the versions.
- Gaming—adding game-like elements to make the AI interactions fun and human-like.

Attempts have been made to classify the new kind of jobs (Wilson et al. 2017), into categories, and sub-categories, such as:

1. AI Trainers: Those who teach/train AI systems
   (a) Customer Language Tone and Meaning Trainer: Trains the AI system to look beyond literal meaning and the superficial to decipher subtleties such as emotional content and hidden meanings
   (b) Smart Machine Interaction Modeler: Mimics humans at work to understand the process and the nuances
   (c) Worldview Trainer: Cultural training, global perspectives

2. AI Explainers: As AI systems grow more intelligent, explainers are needed to deconstruct the system and explain the behavior to business people and consumers in a non-technical manner. The explainers bridge the gap between technologists and businessmen.

(a) Context Designer: Helps make smart, optimal decisions based on different contexts, such as business scenarios, process, and so on

(b) Transparency Analyst: Helps bring transparency into the AI system's thinking process and documents the opaque areas for further improvement

(c) Usefulness Strategist: Helps the AI system regarding when to use what method in specific contexts, for example, when to revert to traditional rules and when to apply heuristics and intelligence

3. AI Sustainers: This is a more operational role, and they ensure that the AI systems work as planned, and they deal with the issues arising out operations.

(a) Automation Ethicist: Evaluates impact other than commercial/economic

(b) Automation Economist: Evaluates the downside risk of substandard performance

(c) Machine Relations Manager: Rates algorithms based on their performance in the field/operations/production

4.2.8   *Immersive Computing: VR, AR, XR, MR, and RR*

Immersive Computing is a technology that blurs the line between physical world and simulated world (Immersive Technology). It is a continuum, or a spectrum as depicted below. VR (Virtual Reality), AR (Augmented Reality), XR (Extended Reality), and MR (Mixed Reality) are all different kinds of Immersive Computing (Joyce 2017). And RR (Real Reality) is the plain physical, real world devoid of any kind of computer imagery or simulation. The spectrum shows RR on the extreme left with 100% reality and 0% computer simulation, whereas VR is on the extreme right with 100% simulation and 0% reality. AR is in the middle.

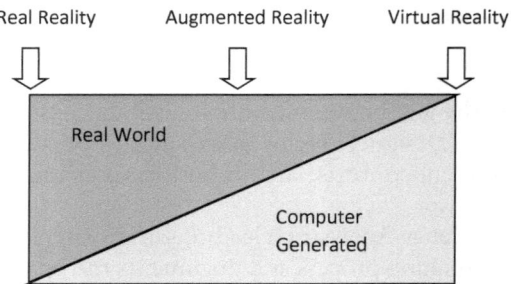

VR (Virtual Reality) is a computer technology that simulates an environment for the wearer of VR headsets, by creating 3D images and sounds such that the user experiences through his senses an alternate computer-simulated reality in a virtual/imaginary environment. The simulation can also give tactile sensations to the user such as vibrations. This is a rapidly evolving field which has many applications. VR is an evolution over 2D simulated environments such as the television or the computer monitor.

Examples of applications of VR include gaming, entertainment, pilot training through aircraft cockpit simulation, military training, telepresence, virtual workspaces, classrooms, tourism, healthcare/medicine, education/training, architecture and construction, and so on. For further reading, refer to VR Brief. Facebook's Oculus is a high-end VR device.

AR mixes information and imagery into the real-world visual/audio/haptic experience for specific purposes. The user is fed input from the real-world environment through direct or indirect views/feeds, and then the desired (augmented) elements are mixed into this reality. For example, the user can watch a tourist monument through his mobile phone's camera, and the AR application can add information about the monument beside the image of the monument. Another example is that of a machine which is being repaired. Watching the machine through an AR application shows the real machine in a live feed, but the AR software also marks the different parts of the machine through superimposed arrows and instructions on how to open the machine and which part to examine. For example, one drawback of e-commerce is the touch-and-feel gap. While browsing goods for personal use, such as clothing, jewelry, sunglasses, spectacles, and so on, the physical presence is missing. AR can bridge this gap, by superimposing the article onto a customer's image, allowing him/her to visualize the experience akin to physically trying it out. This allows customization of tastes, fashion, color, size, fit, and comparison between different products.

AR has wide applicability, for example, e-commerce, gaming, training/ education, virtual showrooms, marketing, travel, and tourism.

In MR (Mixed Reality), the computer simulation/imagery is not merely added into the reality, but the viewer is able to treat them as actual objects and interact with them. Microsoft's HoloLens facilitates MR. For example, play games with objects that are interspersed with reality.

XR or Cross Reality allows a combination of AR, MR, and VR. Google ARCore (earlier known as Tango) is an example of XR and is a platform to build applications or integrate with other applications. In this, the XR application running on a device discovers the device's position and orientation within the environment. The device becomes aware of its environment. It allows applications to be built on this platform to exploit this capability. For example, the game Unity integrates with Tango to allow the user to create or change games by moving the device in the game space.

4.2.9   Gaming

Gaming entered the ICT scene with video games in the 1950s. Since then bulk of gaming applications have been for entertainment. There were some games used for educational purposes, such as storytelling and teaching/training, and in simulations. However, as gaming advanced and became more sophisticated, the academic and scientific community took note and started finding serious applications with gaming (Karasavvas 2017). A problem with traditional approaches to research and development is that of shortage of knowledge and talent. When we use computers as a tool to help us, they are helpful only when programmed. Programming and even Machine Learning and AI have some kinds of boundaries, in the sense they are limited by the models or meta-models conceived by their developers. Gaming helps combine human talent with the power of the machine. And when a large problem is broken into parts, a lot of normal people (known as citizen scientists) can be deployed to help solve the parts of the puzzle, thereby contributing to scientific research and the arts. Instead of letting millions of gamers worldwide waste away their hours with normal games, their human brain "bandwidth" can be utilized for a larger scientific (or artistic) goal. This crowdsourcing is also an application of open innovation.

Examples:

1. The Cure (Good et al. 2014) engages gamers to find diagnostic indicators for breast cancer. It presents a set of genes and criteria in the form of cards, which the gamer has to organize. The higher scoring ensemble wins the game and is a better predictive model for breast cancer. There are similar games for cancer research such as Genes in Space.

2. Second Life (Linden Labs) is an entire world created in VR, in which people can log in and create their own personas (avatars) and interact with one another as in the real world. It is like a massive multiplayer game. It has its own economy and the virtual characters can create their own virtual objects. Second Life is used in technical and scientific research (Lang et al. 2009), and in education, research, and training.

3. Senevirathne et al. (2011) explore how serious gaming is used in sports, healthcare, and exercise.

4. Serious Games International is a UK-based company (SGIL 2018) that builds AR/VR applications for clients for serious purposes—training, leadership development, improving safety in dangerous environments, revolutionizing maintenance operations, and collaborative design in engineering.

5. The book (Cai and Goei 2012) compiles papers on serious gaming applications including rehabilitation, Chinese calligraphy, marine traffic conflict control, fashion simulation, and CNC simulation.

6. The book (Cai et al. 2017) compiles papers on serious games for education. It covers three major domains that use gaming: Science, Technology, Engineering and Mathematics (STEM) education, special needs education, and humanities and social science education.

7. The HIT (Human Interface Technologies) team at the University of Birmingham, UK, is a leader in serious gaming (HIT 2018). They research and develop serious games, in areas of defense and civil medicine and surgery, counter-terrorism, uninhabited air/sea/ground/sub-sea vehicles, culture and heritage, and the environment. Specific examples include interactive trauma trainer, submarine spatial awareness training, unmanned vehicle (land/air) demonstrators, and so on.

Senevirathne et al. (2011) show how six core gaming technologies are needed for all gaming:

A 3D engine to manage all the art assets, user input and rendering visualization, Physics Models to simulate the real world and create a believable environment around the player, Artificial Intelligence to imbue smartness in the simulated game characters to enable them to challenge (or work alongside) the player, Networking to allow multiple players in different locations to play and participate in the game, and Persistent Worlds to allow the game worlds to exist without the active participation of the player.

4.2.10   3D Printing

3D Printing is an additive manufacturing technique in which a machine takes a digital model as input and translates it to reality by combining particles of material (liquid or powder). It is different from other computer-controlled types of manufacturing such as CAM, in that the latter create a product by *removing* material from a block, whereas 3D Printing creates the final product by *adding* material layer by layer. The input files for the 3D Printer are computer/digital models in the stereolithographic (STL) file format. 3D Printing is typically used for rapid prototyping rather than for mass manufacturing.

Thus 3D Printing enables "economies of one" as compared to the older "economies of scale". It is easy and economical to produce one specially designed component or piece. With rapidly changing business models and designs to address the evolving market needs and tastes/expectations of the customers, 3D Printing will gain immense traction. The customer base is also getting increasingly fragmented, and the Long Tail is a real addressable market. 3D Printing is a perfect match to cater to these realities.

Some uses of 3D Printing are:

Special tools, end-of-lifecycle parts, i.e. to make spare parts that are no longer available, customized products, one-off products such as aircraft engines, bridges, machinery, medical devices, prosthetics, replica of sculpture/monument/person.

REDD is one of the largest manufacturers of 3D Printers in India (www.3Ding.in).

4.2.11   Robotics

Robots are computer-controlled machines that are semi- or fully auto-mated, and enhance, extend, or assist human capabilities. Robotics is a complex interdisciplinary field encompassing mechanical engineering, electrical engineering, computer science, mathematics, biology, bio-medical engineering, and so on. Robots are used in a variety of circum-stances from repetitive jobs to exploratory and complex/dangerous tasks. The use of robots is only bound to increase as they can work 24/7, in difficult and dangerous environments, and magnify human capabilities. Robotics and AI go together. The robots will get more connected and intelligent. The uses of Robotics include manufacturing, construction, exploration, surgery, customer service, maintenance, home assistants, defense, and so on.

For further reading, refer to Robotics.

4.2.12   Drones

Drones or UAVs (unmanned aerial vehicles) are essentially unmanned air-crafts, although the term drone can refer to unmanned vehicles on water or underwater as well. They may be remote controlled by a human, or they may be completely autonomous, that is, operated by onboard computers.

Examples of the use of drones are precision farming, aerial survey of disaster-hit areas, disaster relief, defense, autonomous weapons, railway safety, facilities inspection, delivery of e-commerce packages, photography and live media coverage and reporting, surveys especially in difficult ter-rain such as forests/mountains, construction of buildings and roads.

Drones have massive potential, and there are drone regulations being crafted all around the world to prevent misuse of drones, such as in crimi-nal activities, terror attacks, and invasion of privacy.

4.2.13   Blockchain

Blockchain is a relatively new entrant on this list, and it gathered over-whelming attention for being the backbone of the popular bitcoin crypto-currency. This spurred widespread interest in the blockchain technology itself, with the possibility of using it outside the world of cryptocurrency. Implementing a blockchain project needs a strategy (Plansky et al. 2016).

Many industry sectors are now seriously considering putting blockchain to use in different areas. While this seems to be an exciting notion, it is also fraught with risks because the intent of bitcoin was entirely different. Due to the tremendous hype created, many organizations are jumping into the blockchain bandwagon, without examining the real needs or the risks. Hence in the first half of this section, we'll look at blockchain critically, and then move on to the real use-cases.

4.2.13.1   *Blockchain: The Critical View*

Blockchain (especially the public variety) is not meant for all applications. Will blockchain bring the desired innovation and breakthrough in different industries? There are several factors to consider. Blockchain itself is a decentralized ledger that records transactions and resides on a peer-to-peer network through maintaining copies with each of the peers. So, there is no central authority (or authorities) that is a source of truth regarding the transactions; rather the entire peer network collectively arrives at the truth through verification with one another (i.e. cross-verifying with each other's ledgers). A new transaction is added onto the blockchain only when a peer validates it by referring and cross-verifying it with the copies of the blockchain that other peers hold. If there is consensus, then the transaction is added, and the peer who did the validation (mining) gets rewarded, typically with cryptocurrency.

This works well for bitcoin because it wants to avoid the intrusion or even the presence of any kind of authority from anywhere on the globe imposing rules or controlling the transactions. In bitcoin the entire thing is driven by an algorithm, which, among other things, incentivizes the peers on the network to help verify new transactions. For public (i.e. permission-less blockchains) a complex computational process known as mining needs to be done by the peers who compete to decide who is the winner who gets a chance for validating the transaction (and hence collect the cryptocurrency reward at the end). One needs to bear in mind that despite all these conditions, bitcoin has been heavily influenced or disrupted by state and non-state actors. For example, several countries have banned bitcoin, or are treating it as taxable assets. Bitcoin itself has been hacked at least four times. The underlying bitcoin mechanism itself is capable of being overwhelmed and subverted. For example, see how China is controlling the bitcoin mining (Popper 2016).

Due to inherent complexity with the blockchain technology, it has not yet disrupted any industry even after a decade (Stinchcombe 2017). Public blockchain is an extremely energy-intensive process because of the peer-to-peer model and the mining process. For example, bitcoin is 0.01% as efficient as conventional transactions, and if one were to run all of visa credit card's transactions on the blockchain, it would take 5000 nuclear reactors (Stinchcombe 2017). For a single bitcoin transaction, about 9.56 US households can be powered for a day. This is a conservative estimate because some miners around the world (e.g. in China, India, etc.) have found access to free electricity that they're using for mining. Other cryptocurrencies, like Ethereum, can improve the performance severalfold through more efficient blockchain implementations and mining processes, but if the starting point itself is at 0.01% efficiency, then it is not saying much. And the alternate mining mechanisms like proof-of-stake instead of proof-of-work are yet to prove themselves fully in the market.

There are some public blockchain implementations, like that which supports the cryptocurrency Ethereum. These have additional features like smart contracts. In a smart contract, the blockchain software also executes coded business rules whenever the new transaction matches certain conditions, for example, the transaction is above $10,000 or that a certain category of stakeholders did the transaction. These business rules are coded in by one of the parties by consensus from all stakeholders. This seems quite attractive and may work well. However, the flip side always is that the smart contracts do not necessarily carry legal validity. Any legal validity can be achieved only through other agreements among stakeholders that the code in the smart contract is tied to certain clauses of a legal contract residing elsewhere (outside the blockchain). There are also risks associated with software bugs, which causes defective code to be executed, and all the stakeholders are forced to accept the results, that is, output of the software execution. A big problem is that unlike normal, conventional software systems, there is no easy way to upgrade the code in the smart contracts. Once software has been loaded, it is practically considered sealed and locked. Any changes need consensus from all the stakeholders, perhaps through a cumbersome, long process, if at all permitted. Examples of smart contract failures are DAO (decentralized autonomous organization) that is well known (Smart Contracts and the DAO Implosion 2016).

When one is considering using blockchain for use-cases, it warrants asking a few key questions:

1. Is a distributed ledger really needed? Why not just use a centralized database on the cloud, accessible from anywhere, by anyone with security access? The latter option provides a thoroughly solid, well-established, secure, robust, and scalable solution, battle-tested by millions of implementations across the globe.
2. What is the incentive for verification of transactions? In cryptocurrencies, the incentive scheme is to pay a few units of cryptocurrency itself. But for normal applications, what is the incentive?
3. Who will do the mining? Is it a public blockchain? Most organizations would rather stay away from a public blockchain because they see no point in throwing open to public all their business transactions.
4. On the other hand, if the organization decides to use private blockchain, one can do away with the public mining process, and have the transactions controlled by only a few selected parties (such as the organization, its customers, and key stakeholders). Then blockchain is a very convoluted way of doing what a centralized database can accomplish in a far more efficient and secure manner. Where is the real need for using blockchain?
5. Does the organization understand the contractual and legal risks associated with smart contracts? What are the real benefits achieved of avoiding more conventional, but evolving, systems like cloud-based ERP, backed by solid legal contracts? Are the benefits documented and agreed upon by everyone? What about the technical risks associated with smart contracts? If there are software bugs, then what are the contractual and legal agreements among the stakeholders to address them?

If all these questions have been answered satisfactorily, only then may one consider blockchain implementation. For a further list of detailed conditions to check whether a blockchain implementation is really needed, see Greenspan (2015), in which he explores how one can avoid the pointless blockchain project. As per IBM Blockchain, the negative indicators for going forward with a blockchain implementation are:

• Where there are high-performance requirements for transactions: Mining will not suffice. Bitcoin can do at most seven transactions per second.

- When the organization is small and has a limited business network: Where then is the need for such a combination of elaborate verification, transparency, and immutability?
- To attempt a database replacement: Blockchain can never match a database in terms of performance, rigor, and clarity.
- Where a messaging solution is needed: Blockchain was never intended to be used for messaging.
- Where a replacement for transaction processing is needed: Blockchain was designed as a public, distributed ledger on a peer-to-peer network, not a transaction processing engine.

Public blockchain may be suitable for some use-cases like real-estate registrations where verification, immutability, and transparency are all required, as explored in the next section. And these cases are generally far and few, that is, cover an extremely small percentage of the total number of use-cases in the industry. Evidently, the technology is evolving, and many are trying to put it to use in different scenarios, such as banking, healthcare, insurance, logistics, and so on.

4.2.13.2   Blockchain: Genuine Use

From the previous paragraphs, it's clear that blockchain is unsuitable for running heavy applications. It is inefficient in the extreme to use blockchain for data storage or as a transaction processing engine. But what are the genuine use-cases for blockchain then?

The key principle is: *blockchain serves to kill hegemony.*

The typical indicators for a good blockchain use-case are:

1. It is a private (permissioned) blockchain. No mining is needed.
2. Lightweight. Where the complete data does not need to be stored on the blockchain; only minimal information. And there are no high transaction processing needs.
3. There are multiple parties. And they are known parties with proper identities. Usually the parties can join in by invitation or selection only. There are strong criteria for joining.
4. There is total transparency. All the parties have equal rights and well-defined roles regarding the blockchain data and operations. The blockchain ensures that no one party can overpower the other as far as data on the blockchain is concerned. There is no hegemony on the blockchain data and operations, even if some parties

are vastly more powerful than the others, by size, clout, riches, political power, affiliation, location, history, reputation, brand name, sector, type of organization (whether government, private/ public, corporation, NGO), and so on. The data is never controlled by any one party.

5. Need for immutable records, for example, as a source of truth.
6. Where it is a trustless ecosystem. There are no strong reasons to blindly trust any party's records, or one party's over the another's.
7. Where there is a risk that one or more parties may alter/distort the records, often to their advantage and/or to the detriment of the other parties.
8. Where there is a reasonable chance that a good degree of truth can be arrived at in most cases with the collaboration of the parties, that is, cross-verification of data using the blockchain mechanism.
9. Where truth benefits all parties. If not immediately, then over a reasonable period of time. That is, the net benefits of truth vastly outweigh the sum-total cost of distortions of truth. In other words, all the collaborating parties find it simpler and more inexpensive to adhere to the truth. And distortions by any of the collaborating parties are exceptions rather than the norm.
10. Where there is a reasonable recourse to the law in case of deviations. That is, there is an incentive to adhere to the truth, through regulations and legal contracts. There is a strong legal and contractual basis to everything that the blockchain does, that is, the operations/transactions/data.
11. Where the benefits of collaborating overwhelmingly outweigh the alternatives, that is, there is a strong incentive for joining or remaining in the collaboration.
12. Eliminate middlemen and brokers of all sorts.
13. Eliminate the need for audits.
14. The data on the blockchain should be accessible with a public/ private key-pair and based on the conditions/filters put for the specific request. For example, if one employee within a hospital is responsible for feeding a patient record into blockchain, there is no need for him to access the entire medical history of the patient that the blockchain holds. Similarly, an applicant may decide to reveal only the past five years of his medical history to an insurer.

Provisions need to be made on the blockchain to enable such role- and permission-based access as needed.

15. All agree to abide by the rules of the blockchain. If there are smart contracts (i.e. rules embedded in the blockchain in the form of software code), then all agree to abide by the software's output. Hence there is additional responsibility for each of the parties to verify and validate the software before deploying it on the blockchain. Each may employ their own means (including outsourcing) to test the software independently. If there are bugs or problems discovered, then there should be commonly agreed mechanisms in place to raise complaints and rectify the software. There is always a small risk that the bugs will cause damage (financial or otherwise) to some parties. But the benefits far outweigh such risks. Hence the parties should settle this amicably.

An example will make it clear.

Banks in India face problems when issuing loans. NPAs (non-performing assets) or bad (irrecoverable) loans are common. There are also frauds of all sorts, such as:

- In the trade finance sector, businesses present the same merchant discount bill (with the same invoice number) to multiple banks.
- Loan applicants take loan against the same asset (collateral) with multiple banks.
- Loan applicants with bad credit history with one bank manage to get a loan with another bank.

Not only does the mandatory KYC (know your customer) process need to be made foolproof, but there is a need for KYE (know your employee). The recent major frauds in Punjab National Bank (Miglani 2018; Money Control 2018) to the tune of US$3 billion revealed a collusion among the bank employees to run major fraud undetected for years.

In February 2017, a group of 34 banks formed an industry consortium called Bankchain. They implemented blockchain software to share loans-related information among them. For example, they can now verify whether a loan applicant has any bad credit history with any of the other 33 banks, or whether he has taken any other loan against the asset mentioned in this new application form made to this bank. Hence, they can eliminate fraud for many common scenarios pertaining to the members in the consortium. The risk exposure is restricted to unknowns outside the consortium. For example, if an applicant has bad credit history in some other bank outside the consortium, then the risk is contained to that level. The risk is progressively minimized as the consortium expands, and as they build more scenarios into the blockchain, and as they mutually share more data. If there are smart contracts, it will provide greater benefits as fraud becomes very difficult with many rules being coded, loopholes being plugged, and awareness increasing among the banks about the applicants and the possible scenarios.

Is this a fit use-case for blockchain? Applying the indicators mentioned above, the analysis is as follows, against each of the indicator.

Indicator	Brief	How the example matches the indicator
#1	Whether private?	Yes, private and restricted to the members of the banking consortium. No mining needed.
#2	Lightweight?	Yes. Only the loan details need to be stored, such as ID of loan applicant, date, amount, terms of the loan, collateral description, and so on. The loan documents, or the collateral's documents, need not be stored. The information is lightweight, and there are no heavy transactional needs.
#3	Multiple parties?	Yes. The multiple parties are all the members of the consortium, that is, the 34 banks as on date. Strongly identified.
#4	Equality and transparency among parties	Yes. None of the bank members of Bankchain has a higher priority, regardless of whether it is a larger or smaller bank. There is total transparency. Nobody gets more access.
#5	Immutable data?	The loan records should not be tampered with. This is ensured with the blockchain's distributed ledger. So, if a fraudulent loan applicant has a friend inside a particular bank who will help tamper the applicant's past loan record to enable him to get a loan, this scheme will not work, because they will have to tamper with the records in the distributed ledgers of all 33 banks to get his way. This is a near impossible task. So, the records are immutable.
#6	Trustless?	Yes. It is a trustless network. Hence verification of identities and protocols are essential. Can't blindly trust the records from any one bank member, because its data copy may be obsolete and not updated. Chances that before submission, the information may have been compromised, or withheld, by vested interests during internal processes.
#7	Equal trust?	Yes. The data from all banks in the consortium is treated with the same trust as far as blockchain is concerned.

(continued)

(continued)

Indicator	Brief	How the example matches the indicator
#8	Risk of tampering?	Yes. There is a constant risk that any bank may be subject to hacking or tampering, for example, because some employee inside the bank may have morphed some records in connivance with others, or that there are loopholes of all sorts.
#9	Converges to truth?	Yes. By verifying the loan history and collaterals of an applicant based on the data from all the banks in the consortium, a bank can get a good truthful assessment about an applicant. The blockchain mechanism helps converge the parties towards truth.
#10	Good benefit/ cost ratio, collectively?	Yes. The loan information that is shared on Bankchain benefits all the banks in the consortium. They get much more than they've invested in it. Currently in India 5% of the loans are NPAs (non-performing assets) or bad debts. This can be cut down.
#11	Legal?	Yes. The bank has every legal right to reject a loan on the basis of information gathered from the other banks (in order to detect fraud). Recovery of loans and seizing personal assets of a defaulter are all within the purview of law.
#12	Individual benefit?	Yes. As the use-case scenarios expand and deepen in Bankchain, and more banks join Bankchain to collaborate, then the value that the banks get out of each other's loan information is far higher than going solo. Currently each bank gets the loan data of 33 other banks from the consortium, whereas it just submits its own loan information, and pays membership fees for Bankchain.
#13	No middlemen?	Yes. The bank need not employ any middlemen or brokers who will supply and verify loan applicant-related information for the bank. It's all available on Bankchain's blockchain.
#14	No auditing?	The Bankchain is self-auditing due to the nature of blockchain, in which all the parties hold copies of past loan information. The blockchain does automated auditing by simply cross-verifying information with the records of the other banks. The role of auditors is minimized or eliminated altogether.
#15	Filtered querying?	Any query on Bankchain can be built to be restricted by design in several ways, for example, one of Bankchain member's queries about a specific loan applicant's history can be restricted to just that in the API call. There is no need to allow the bank to pull out history of some other borrowers of the other banks. Through public/private access key sharing mechanisms, both security and privacy of information can be achieved.

(*continued*)

(continued)

Indicator	Brief	How the example matches the indicator
#16	Compliance to blockchain rules?	The banks in Bankchain abide by the rules for feeding in or retrieving data. With any complex scenarios coded in the smart contracts, each bank in Bankchain may get the software tested independently. Problems resulting due to bugs may cause a few bad decisions in stray cases, such as a loan being given to a fraudulent applicant, and so on. Barring such corner cases, the system should work well and yield immense benefit most of the time. Rectification mechanisms should be in place.

Hence it can be concluded that Bankchain is a good case for implementing blockchain.

Similar examples exist in healthcare industry, where blockchain implementation is used to capture medical records and transaction history of patients. This information is made available to health insurers, hospitals, diagnostic centers, and doctors. Many processes are simplified, such as insurance claims settlement, insurance application processing, and even treatment because the doctors have ready access to the patient's medical history. The patient is empowered because his medical records follow him, rather than being locked up in one hospital's proprietary database. He gets the freedom to switch hospitals, take second opinions, get access to his own medical history, right from diagnoses done, treatments undergone, tests taken, and medicines bought/used. He can then give selective access to any party on the blockchain, for specific purposes.

Another example is that of real-estate transactions. This is what the Telangana state government in India is embarking on. In a real-estate deal, the transacting parties need to be authenticated. Once a deal is sealed and ownership transferred, it needs to be etched in as a permanent record that is immutable. The actual transaction details constitute minimal data. The documentation need not be kept on blockchain. One asset should not be sold or mortgaged to multiple parties at the same time, which constitutes fraud. Unlike e-commerce or regular banking, there won't be millions of transactions on an hourly or daily basis. Hence there is no danger of the blockchain becoming an intensive transaction processing engine.

In the above cases, again, the requirements are that the data be tamper-proof (immutable), lightweight, transparent, accessible to all parties, with equality among them, is legal, yields benefit, and has consensus among all parties.

4.3   THE INTERNET

The Internet itself deserves a closer examination. There is no doubt that the Internet is one of the greatest technological inventions ever, and it has revolutionized the way the world communicates and functions. From corporations, governments, banking, healthcare, and infrastructure, to educational institutes, social media, entertainment, all depend on the Internet for needs ranging from the critical to the informational, right down to the trivial. However, the Internet's architecture is aging rapidly. This subsection is meant to highlight the problems and allow you to consider options where applicable. Be cautioned that parts of this sub-section are a bit technical. Feel free to take only the gist of any pieces that you find too involved. Read on until the end, where Web 3.0 has been briefly discussed.

The expectations and demands imposed on the Internet have caused it to reach a breaking point. It has been burdened far beyond the limits of its original design and construction, in every conceivable manner:

1. *IP Address Space Limitation*: An explosion in a number of deployed nodes, namely, computers, mobile devices, sensors for IoT, televisions, gaming devices, and so on, means that the Internet needs to cater to over a 100 billion IP addresses, which the current IPv4 addressing scheme cannot accommodate by a long shot. The shift from the 32-bit IPv4 to the 128-bit IPv6 address will allow far more IP addresses, but will introduce security problems of its own (IPv6 Security Issues). It can also cause sophisticated network problems that are very difficult to diagnose. Currently there is a period of hybrid existence, where both IPv4 and IPv6 co-exist and this compounds the problem (IPv6 Migration Problems).

2. *Piling on Usage* through different means such FTTH, Mobile 2G/3G/4G/LTE (and now 5G), traditional landline Internet, and so on.

3. *Explosion in Different Types of Usage*—video-on-demand, chatbots, social media, IoT, mission-critical real-time needs, cloud, e-commerce, moving entire businesses online, mobile apps, and so on.

4. *Planet-Scale Implementations and Always-On Culture*: The Internet has moved from its original philosophy of "Connect when you need to" to vastly different expectations of:

"Always-On", "24/7 Availability", "Minimum Expected Bandwidth", "Performance graph should scale Linearly with the Load, even with mixed typed of work-loads", "Upload speed is as important as Download speed, because of Social Media, P2P, Video-Sharing", and so on.

The infrastructure is strained to the point of collapse under the pressure of all these expectations. Many cloud-based implementations of applications are now on planet-scale, with expectations of zero downtime and response time in milliseconds across thousands of nodes serving millions of users. Examples not only include Uber, Alibaba, Facebook, Amazon, and Google but also most corporate and government services that are moving online, such as telecom services, webmail services, government income tax sites, HR service providers, logistics, airlines, travel booking, hospitals' self-service portal for patients, and so on.

5. *Regulatory Requirements*: Service providers face a variety of challenges ranging from net neutrality to anti-monopoly acts, spectrum auctions, and bandwidth sharing agreements. Currently the Internet is deemed a utility that should be always available on-demand (24/7), and the right to Internet access is a fundamental right for the citizens.

6. *Dominance and Control by a Few and Privacy/Security Concerns*: Although the Internet was originally conceived as a decentralized, distributed, and self-organized network, it is now dominated by a few large corporations such as Google, Facebook, and Amazon. And there is severe loss of citizens' privacy owing to data gathering by these corporations and surveillance by governments. A handful of people can shape everything right from public opinion and tastes to purchasing habits of people. The Internet has essentially become an advertising platform. All your data is easy prey. On the dark side, unknown parties can hack/steal/manipulate information, resources, money, rig elections, cause countries to go to war, sell drugs/weapons, destroy reputations, indulge in crime, and cause large-scale disruptions by operating from behind the screen in unknown ways. The Dark Net is the underbelly of the Internet, and in fact much larger than the visible net. Even protecting one's PCs, mobile devices, and servers has become a major challenge with lurking hordes of new and evolving malware/viruses/bots/hacks/exploits of every conceivable kind.

Other than being loaded to its limit, the Internet also has the following fundamental, inherent limitations (Internet's Fundamental Limitations 2010). For those technically inclined, here is a quick point-wise summary:

1. *Processing and Handling Limitations:*
   a. Difficulty in diagnosing potential problems, little feedback from network for hosts to perform RCA (root cause analysis), vulnerability to different kinds of attacks, and so on.
   b. Missing data identity in the network, deficiency of methods for dependable, secure, global processing and handling of network/ systems infrastructure and basic and critical services, lack of essential services for critical needs such as regulatory compliance, healthcare, disaster warning, and so on.
2. *Storage Limitations:* Lack efficient means of storage, lack of inherited data integrity, reliability and trust, inefficient caching, and mirroring.
3. *Control Limitations:* Lack of flexibility in control components, data, and control separation mechanism is much too outdated and inflexible.
4. *Transmission Limitations:* Inefficient transmission of content-oriented traffic, poor/missing inherent security mechanism in the communication infrastructure, lack of unified architecture of IP control plane, inefficient congestion control.
5. *Operational Limitations:* Some network segments lack bandwidth capacity leading to high costs being incurred by communication service providers, without seeing any business ROI. Inter-domain routing systems are peaking out in limits. Elastic scaling capability is missing, especially during sudden spikes in traffic. Inability to handle the new demands and challenges of exponential growth in information, in a fine-grained manner, especially in ranking/classifying/filtering/retrieving information. The current Internet architecture makes it extremely cumbersome and expensive to access, retrieve, and integrate information from resources across the Internet.

The growth of the Internet is unsustainable with its current architecture, the implementation, and its ever-increasing load. Problems are proliferating by the day, for example, security nightmares, bandwidth problems, breakdowns, ever-increasing complexity, and shortage of expertise.

All these reasons have led to the rise of alternatives to the Internet. These will not replace the current Internet overnight, but expect these to grow in implementation and adoption over the years:

- *IPFS*: InterPlanetary File System is an open-source, peer-to-peer distributed file system that connects all the users to the same file system. It does this through a hypermedia protocol designed to create a faster, secure, and more open web. It implements a protocol that holds a content-addressable, P2P technique of storing/sharing hypermedia in a distributed file system. IPFS does away with the Internet's centralized DNS servers. Content can be stored on hundreds of thousands of nodes across the network and be accessed directly, efficiently, and securely without going through a hierarchy of centralized servers. It replaces the HTTP with its own IPFS for addressing and accessing content. For further information on IPFS, read IPFS Resources.
- *Wireless Community Networks*: This is a grassroots approach to building a flexible network architecture, to provide free or subsidized access to the Internet. It attempts to replace the (commercial) access to the Internet provided by ISPs. Such an approach also facilitates local interactions bypassing the Internet altogether and fosters community building and participatory design (Wireless Community Networks). The types of design solutions include Cluster, Mesh, WISP, WUG, or Wireless Users Group. An example of Wireless Community Network is AWMN (Athens Wireless Metropolitan Network). For a longer list of examples, see Wireless Community Networks Implementations.
- *Peer-to-Peer Systems and Encryption*: These enable the distributed implementation of computing services like retrieval, storage, content distribution through partitioning the tasks and workloads between peers on the network. Successful applications include BitTorrent and Gnutella for file sharing, bitcoin for cyber currencies. For further examples, see P2P.
- *Distributed Web-Based Social Networks*: This is a kind of social network that is decentralized and distributed across multiple distinct service providers. Social media then becomes a public utility that has many different social networking sites, and each site can connect with the

users of the other sites. Users can build their own sites and organize them as per their own preferences and set their own levels of privacy, filtering, and security. Examples are Diaspora and Ello. Organizations (especially governments and corporations), or individuals who do not wish their data to be mined by the popular/commercial social networks, may adopt distributed web-based social networks.

For a longer list of alternatives to the Internet, see Alternatives to Internet. These alternatives provide some combination of the above, following principles such as P2P, mesh networking, decentralized systems, inbuilt cryptography/security/privacy/anonymity, robust protocols, addressable content, and so on.

However, there are no easy alternatives. Any options will need to be built slowly over time. Until then, cybersecurity, efficiency, and privacy on the Internet will continue to remain a challenge, and every individual and organization will need to find its own ways of dealing with these factors, depending on the unique requirements and available budgets. Communication service providers, governments, and corporations may choose to implement and provide alternatives to cater to different needs. For instance, it may be far cheaper to provide a WUG locally rather than full-blown Internet access. Similarly, disaster warning systems, or sensor infrastructure, may deliberately be isolated from the Internet to avoid hacking or outage due to a DNS failure. To cut down the huge traffic generated by sensors for an IoT network, a lot of computation can be done locally, for example, Cisco's edge router that aggregates data from sensors and takes local decisions intelligently. It also shields the vulnerable sensors from cyber-attacks.

Web 3.0 is a new version of the Internet being conceived which will be pro-privacy and anti-monopoly (Zago 2018). Web 3.0 offers many advantages such as no central point of control made possible with the underlying blockchain platform, ownership of data by the end-users who can encrypt and control access to it, elimination of most data breaches/hacking, high interoperability among devices/software and easy customization, 24/7 uninterrupted service due to removal of single points of failure, and permission-less blockchains with no controlling authorities anywhere and free access to resources anywhere regardless of geography, citizenship, and so on. DAPPs or distributed apps will be Web 3.0 equivalent of Web 2.0 Apps.

Examples of the Web 3.0 applications:

Application	Web 2.0 App	Web 3.0 DAPP
Storage	Google Drive or Dropbox	IPFS, Siacoin, Storj, Casper
Video and audio calls	Skype, Yahoo Video Chat, Google Meet	Experty.io, Xmoneta
Operating system	Android, iOS, Linux	EOS, Essentia.one, AargonOS, Brahma OS
Messaging	Google Hangouts, WhatsApp,	Status, Lisk, EtherTweet
Social network	Facebook, Twitter	Akasha, Steemit, Leeroy, Vevue
Browser	Google Chrome, Firefox, MS Edge, Apple Safari	Brave, Toshi, Moon, Token

For more examples of DApps, visit the site:

- State of the DAPPS (https://www.stateofthedapps.com)
- Fueled.com (https://fueled.com/blog/dapp-development-companies)

4.4   DECONSTRUCTING THE CTO/CIO ROLE

In a world where software and algorithms run most parts of our lives, ICT is too important to be a mere department; hence it needs representation at the C-Level in every organization. The exact role of such a chosen head who runs the ICT function in a company needs elaboration.

A CTO (Chief Technology Officer) typically runs the product development organization, whereas the CIO (Chief Information Officer) runs the internal ICT organization. A CIO heads the organization that runs the operations and builds reliable and repeatable processes. A CTO serves as the company's top technology architect and runs the engineering group, using technology to enhance the company's product offerings (TechRepublic CIO/CTO). The lines between the roles of CIO and CTO have blurred and vanished; hence many enterprises have now combined the role into a CTIO (or CITO) role. The key priorities of the CTIO are:

#	CTIO priority	Purpose	How it is accomplished
1	Customer focus and customer experience	Raise/improve the topline	*Aligning ICT to business*, digital transformation, conducting proof-of-concepts
2	Budget control	Manage/control bottom line	Cost management

(*continued*)

(continued)

# CTIO priority	Purpose	How it is accomplished
3 Innovation	Create new business value	Build new (or improved) products, solutions, models, processes
4 Performance	Run operations	Automation, reusable processes, frameworks, standards compliance
5 Growth	Provide greater business value and handle larger customer base	Scale, optimization, integration, data sciences, analytics/mining, insights

Thus, the CTIO's role and job description includes the following:

Value Driver, Strategic Thinker, Delivers Business Value, Showcases Products/Solutions to the Customer, Leads Product Management and Product/Solution Engineering, collaborates with the Business Units and with the CEO/COO/CMO/CSO/CXO, Guides the board, Influences Strategy, Recommends New Technologies, Runs ICT Operations

Over the years, the role of ICT has shifted from that of carrying a support responsibility into a core strategic function that drives and catalyzes the entire organization. Hence *aligning ICT to business is the topmost goal* for a CTIO. What does ICT mean for different stakeholders?

Stakeholder's point of view	How ICT aligns to business	Examples
CEO (Chief Executive Officer) and the Board	ICT is right at the core of strategy and governance	Business intelligence, systems and business rules that align with the strategic vision, analytics
Business units	ICT provides agility. ICT is an enabler, catalyst, and differentiator	Cloud, business-specific software, data sciences for insights
CFO (Chief Financial Officer)	ROI, insights and investment decisions, cost reduction	ERP—finance, budget, and ROI calculators
COO (Chief Operations Officer)	Security, risk and compliance, stability/scale/predictability of operations, optimizing resources, procurement, project management	IaaS, SOC (security operations center), disaster recovery for business continuity, opex models, IoT for smart operations

(*continued*)

(continued)

Stakeholder's point of view	How ICT aligns to business	Examples
CSO (Chief Sales Officer)	Customer relationship management, deal making, deal management, order management, account management, sales pipeline management	CRM, order management, concept-to-cash flows, bid management
CMO (Chief Marketing Officer)	Industry analysis, competitive analysis, market positioning, go-to-market strategy, digital marketing, proof-of-concept, marketing collateral	Market intelligence databases, rapid prototyping, competitor analysis, scenario building, and "what-if" analysis tools
CXO (Chief Experience Officer)	Help desk and support system, customer experience, analytics, and feedback into product management	Trouble ticketing, troubleshooting, data mining, UI/UX design
Legal	Contracts, compliance	Contract management systems
HR	Recruitment, performance appraisals, compensation management	HR management systems

Traditionally, successful CTIOs spent their time on these dozen areas:

1. Identifying users and applications
2. Showcasing products/solutions/proof-of-concepts
3. Requirements engineering and product management, of which identifying stakeholders' needs and use-cases/user journeys is a critical part, managing expectations
4. Enterprise architecture, product/solution design
5. Recruiting top talent
6. Technology evaluation and selection, techno-commercial analysis
7. Budgeting, ROI/TCO/benefit-cost analysis, make-or-buy decisions, in-housing vs. outsourcing, procurement
8. Audits of technology environment and processes, review mechanism, and situation analysis
9. Vendor/partner management, vendor analysis
10. Planning and processes, data sciences, analytics, mining, and insights
11. Studying and training technology and industry landscapes and the upcoming trends, networking, and attending conferences and industry meets
12. Research, innovation, experimentation, prototyping

The above list still holds, though the mix varies by organization, by phase, and the exact needs. However, the ground is shifting; there are new concerns and opportunities.

4.5   THE FUTURE DEMANDS OF THE CTIO ROLE

With changes in the industry, technology, regulations, and competition, there are bound to be major changes coming up in the CTIO role. Here are the major demands and challenges that the CTIOs will face in the coming years.

4.5.1   Customer Engagement

Due to the exponential pace of advances in technology, the CTIO is expected to be on top of the technology trends relative to the business in question. The future CTIO needs to understand the technology as well as the business also thoroughly in order to make the right judgment at the right time. Great organizations shape their technology around their business needs. The CTIO will need to meet the current/dynamic customer demands as well as future proof the business.

4.5.2   War for Talent

With rapid advances in technology, finding and retaining top talent will remain the foremost challenge for organizations. Paradoxically, while technology and automation cause redundancy and layoffs, finding the right talent at the right time will remain the biggest challenge. The lines from the "Rime of the Ancient Mariner", "Water, Water everywhere/Nor any drop to drink", have never rung truer before. To say that it is a war for talent would still be an understatement. Far from being a HR-owned problem, it will become the entire organization's problem, because of so many dimensions that go into hiring somebody—the detailed needs analysis, job description, budgets, urgency, the impact analysis if hiring is delayed (or doesn't happen), fitment, duration of hire, joining time, contract/full-time, availability of experts for the interview panel, the hiring process itself including levels of approvals.

Example situations are:

- Need an IoT platform expert for an IoT project that is commencing in three months.
- Our blockchain project has run into problems, and we need someone urgently to troubleshoot the design and implementation.
- A budget of $500 million has been sanctioned by the management of the global company for transforming the entire organization through ICT, over three years. Need an enterprise architect who will drive the transformation and create the roadmap first. Need to hunt worldwide for such a high-profile architect.
- We're holding major events in line to kick off our new sales strategy. Our digital marketing campaigns need to take off well in time for the events so that enough people register for the events. Need to build a team of UI/UX and web-design experts very fast, who will work with our outsourcing partners for web development.

Given the *near impossible situation*, the hiring strategies also need to be innovative and effective. Budget allocation for hiring also needs to be a bit long term to gain flexibility. The challenges lie not only in finding the right talent but also in retaining the existing talent pool.

Examples of such challenges are:

Regrettable attrition due to employees seeking career advancement elsewhere, scarce talent being too expensive, experts retiring, talented consultants who are pulled off in the middle of the projects for other work, consultants whose contracts have expired, overseas consultants with visa renewal problems, managers leaving the company and taking good team resources with them, decline in productivity of an employee who is demotivated for some reason (personal/workload/manager/co-workers), contractual terms and conditions restricting utilization of resources to their full capacity, org resources with outdated skills, and so on.

The gig economy will solve some problems while piling on some others. There will be a plethora of choice and engagement models, such as vendors that are:

Small orgs, large orgs, freelancers, onsite/offsite/online consultants, fixed price, daily/hourly contracts, outcome-based compensation, joint partners, profit sharing, and so on.

This is good, but there are huge associated burdens like vetting and verification, contractual complexity, legal problems, arranging for payments at the right time, measuring and monitoring work constantly, and preventing corruption and kickbacks.

The fragmented and dynamic nature of engagements in the gig economy renders the entire program highly complex and risky. This includes every single area/sub-area one can conceive of: planning, execution, delivery, support, sales, project management, budgeting, engineering, quality control, delivery, marketing, sales, compliance, and so on.

The simplicity has vanished. And this is the reality that CTIOs need to live with. The entire organization will be in the throes of such complexity and will look to the CTIOs to provide ICT platforms, tools, databases of information, and standards to help everyone manage all this.

4.5.3   Requirements Engineering and Product Management

This will remain challenging, more than ever before. The customer/user needs will be dynamic and demanding. There is much to compare with in the industry. Due to obsolescence, the implemented solution will always fall behind the curve, while the users will expect the solution to rival the latest and greatest offering in the market at the most minimum cost (even free if possible). Unearthing and showcasing value will be critical. And this will go together with prioritizing requirements, and product management. Read the chapter on focus for details on how to get this right.

4.5.4   Getting the Enterprise Architecture Right

With changes happening in every part and every layer of the complex ICT system sprawling the organization, it is easy for sub-systems to go out of sync. The enterprise architecture ensures that the organization continues to deliver high value with minimal wastage. The Open Group Enterprise Architecture Framework (TOGAF 2011) is the de facto global standard for enterprise architecture. It explains very well what an EA should be and what it should do. There are excellent industry frameworks too, such as Frameworx (TMForum Frameworx). Getting the enterprise architecture right saves the organization of many problems that typically occur down the line. The Frameworx principles are also excellent to study and use.

For example, the architecture could address multi-cloud deployments, IoT, adoption of AI, integration with legacy systems, and so on.

4.5.5   Integration

Closely associated with EA is integration. With the plethora of (constantly evolving) tools and systems available today, fragmentation is rife. There is high risk of building poorly integrated systems that are somehow cobbled together, leading to unintegrated silos, a spaghetti-like architecture, resulting in dysfunction in many areas. The entire architecture, design, and implementation will be brittle, leading to breakages when changes are made in any part. It will be extremely expensive to maintain or live with.

The worst hit will be the data, which will get fragmented and scattered. Hence it will be very difficult to do reports or analytics or get any insights. The systems and their processes should be integrated ground up, by following a comprehensive enterprise architecture. It is important to have a unified architecture and also follow open standards, to avoid a terrible vendor lock-in. The architectural modules should be loosely (not tightly) coupled.

4.5.6   Data

Due to the sheer scale and complexity of data that will power products, the CTIO will need to create a vision of how to acquire, use, and monetize data. There is an ever-increasing number of data sources, for example, web, mobile, IoT, kiosks, customer contact centers, and so on. And there are too many data solutions to choose from. The technologies are also evolving at a blinding pace. The CTIO needs to keep abreast of all these developments, especially the cutting-edge technologies like AI, Machine Learning, and IoT, and choose solutions beyond the ordinary which will put the organization in the forefront in the industry.

4.5.7   Agility

The modern organization needs agility. Inertia is anathema to the future organization, but is precisely what will set in, with increasing complexity in the organization. Getting things done will neither be simple nor be fast, as systems and data grow. Entropy sets in. With increasing digitalization, the CTIO needs to help shorten both the decision-making process and the time-lag between decision and action. This needs adoption of agile processes, usage efficient architecture/design, cross-functional coordination, and removal of inefficiencies. A common problem is analysis paralysis, which the CTIO should guard against. Prioritization is key; otherwise time and resources are wasted on solving smaller problems.

4.5.8   Security

This has been elaborated in other chapters. Providing security will be the biggest nightmare of a CTIO. The entire business and its reputation rest on systems, data, and the people. Security will neither be easy nor permanent. It will be a 24/7 challenge and ever evolving to take on new threats. The factors that complicate security are borderless networks, security skill gaps, BYOD (Bring Your Own Device), contract staff, changing technologies, imperfect integrations/upgrades/configuration, legacy systems, evolving malware, zero-day exploits, insider attacks, securing IoT, procedural gaps/ loopholes, lack of awareness in staff, lack of compliance despite awareness/ training, and so on. The CTIO needs to shape a comprehensive InfoSec strategy and keep it updated. An ever-vigilant security task force must be constituted which will appropriately address the myriad security threats.

4.5.9   Policy

The CTIO must roll up all of the above into policies for the organization. This will include policies to address regulations, security, architecture, ICT development, ICT operations, data sharing, vendor management, customer engagement, product management, UI/UX, reports, training/ learning, and research. In this manner, there will be no ambiguity for the employees, vendors, software developers, system administrators, and business units as regards how to go about conducting themselves. They will also have a clear roadmap as to when they are getting their deliveries. For example, vendor management could stipulate that the ICT organization partner with two large vendors and two smaller vendors for sourcing talent on contract (consultants). The rationale could be that two of each will keep a healthy competition going and the costs realistic. And use smaller vendors to fill any urgent, small gaps, as larger vendors may not be able to do so.

4.5.10   Training and Development

The blitzkrieg pace of technology and business means that people will fall out of pace very fast. It is not enough to hire the best talent, develop solutions, and build policies. All the people need to be brought on board with the latest developments as regards their work. It is likely that their work itself may be evolving due to the changes. Hence training and development is no longer the purview of just the human resources department. As with other things, it will be heavily driven by ICT. In fact, a lot of trainings will be about ICT systems. Hence the CTIO needs to build a strategic plan for training and development. It should cover online help, technical documents, requirements, business workflows, FAQs, online training, videos, AR/VR simulations, classroom training, and so on. Certifications should be issued. The organization should keep a keen track of the level and availability of talent/skills available in the organization. Training should be part of the checklists, for example, when a new ICT system is released for the organization, the delivery should be succeeded by multiple training sessions. And the recorded sessions should be available online, so that people can watch at their own convenience and pace.

4.5.11   Innovation

This rounds up the list. Innovation will be a prime focus of the CTIO in the new world. Much has been discussed in the other chapters of the book. One cannot survive in the new age without ICT-driven innovation. Open innovation is a given in organizations; the differentiator is how well the open innovation is executed.

The Takeaway Box: Recommendations for the CTIO

- First, thoroughly map the current capability of the org, in terms of ICT talents and skills.
- Understand the vision and roadmap of the organization and work with the key people in the organization to map the ICT skills needed in the *near future* (with a look ahead of a couple of years).
- Plan a little long term and get block budgets. Then hire whenever you can, and whenever you find the right talent; do not wait until it is a desperate situation. The cost burden of this approach is minor as compared to the risk of losing projects, failed products/ projects, losing customers, and the damage to the brand name of the organization.
- Develop ICT policies appropriate to your needs, capabilities, and size of your organization.

4.6   The ICT Spend

It is evident that ICT generates disproportionately great business value, far beyond what it costs, and recurring for long after its deployment. But what do ICT expenditures typically consist of? It is essential to get a good sense of what these are.

Currently, on an average, organizations invest about 4–6% of their revenue in ICT (*CIO* magazine). A study of companies in the USA by Alinean Consulting found that the small and mid-sized companies spent more on ICT (6.9% of revenue) as compared to larger companies (3.2%) (*CIO* magazine). The telecom industry is an anomaly because it is a technology engine in itself, and Telcos may spend 15% of their annual revenue on its network maintenance alone (Microsoft Blog on Telecom Operators 2013). All these figures indicate the status quo. Evidently as the other functions in the organization get digitalized and automated, the figure is only bound to increase, for example, the digital marketing component of marketing, website content creation for all departments, project management tools for programs, field force automation tools for sales people in the field, and so on.

The broad heads under which ICT investments, projected for 2018, are distributed are listed below (SpiceWorks 2017). These are for US/Europe companies, but are indicative figures:

#	Heads	Percentage of overall ICT cost
1	Hardware	31%
2	Software	26%
3	Hosted and cloud-based services	21%
4	Managed services	15%

Source: SpiceWorks (2017)

The breakup is contingent on the size of the company but doesn't vary too much. The managed services portion increases with the size of the company (from 10% for small companies to 20% for large ones).

Here is a further breakup of hardware and software:

Sub-head under hardware budget	*Breakup %*
Desktops	17
Laptops	15
Servers	13
Networking	8
Others (tablets/mobile devices 7%, security appliances 7%, power and climate 6%, printers 5%, external storage 5%, telephony 5%, peripherals 4%, miscellaneous 9%)	47
Total	**100%**

Source: SpiceWorks (2017)

Sub-head under software budget	*Breakup %*
Operating systems	11
Security software	10
Productivity software	10
Virtualization	9
Industry-specific apps	9
Others (business support apps 7%, backup/disaster recovery 7%, database management 7%, IT management 6%, email servers 5%, communications 4%, developer tools 3%, miscellaneous 11%)	51
Total	**100%**

Source: SpiceWorks (2017)

The above table lists the *average* breakups. However, the breakup varies from small to large companies. The smaller companies spend more on online backup/recovery (15%), productivity solutions (10%), email hosting (9%), and web hosting (9%), whereas large companies spend more on virtualization (12%), databases (9%), and operating systems (9%) (SpiceWorks 2017).

For a sample of breakup of IT budgets for healthcare providers, refer to Gartner (2017, 2018).

4.7   CONCLUSION

It is all important to keep tabs on the ever-evolving technology and enterprise landscape, so that one can leverage it well before the competition does. Far more business models have sprung up owing to the developments in underlying technology than the other way round (i.e. technologies developing as a result of business models). To drive home this point, consider the opposite. It would be futile to attempt to build an Uber-like business using the technologies of telegraph and horse cart. Development of a usable technology precedes commercialization and development of standards. Tourism developed because of airlines/railways/automobiles, media and entertainment increased a millionfold due to satellite communication/Internet/consumer electronics, and practically every sector exploded because of developments of ICT such as microprocessors, networks, storage, and software platforms.

Another way to look at it is that the revenues from businesses built based on a technology easily dwarf the revenues got from the direct sale of these technologies. Examples include open-source technologies like Linux, mobile technologies, PCs/desktops, software like databases, ERP, and so on, all of which cost far less to build and sell than the business value they unlocked. In fact it is not worth building and selling a new technology unless the total business impact due to the cascading effects of these technologies is far greater than the direct sales proceeds of the technology itself. Businesses demand a good ROI for technologies from the vendors.

Thus, a successful technology is really an ecosystem enabler, not an end in itself. This chapter described the key ICT technologies that will play a major role in paving the road to the future. It also shows briefly what else it takes for these technologies to enable business. It takes a diverse set to make this happen: trained manpower, leadership, strategy, innovation, customer engagement, management, especially financial management, policy, and ecosystem.

To summarize the main impact of technological revolution, we can argue that it changes completely the way we used to conduct business traditionally. Every organization needs to feel, think, customize, and use technology with a goal and mission in hand to face the ICT Tsunami. The following table provides a snapshot of the main features which have changed the business landscape.

Feel technology differently	Customize technology for operation	Use technology for economic benefits	Think technology for future strategy
Sense	Efficiency	Consolidation	Data
Control	Speed	Differentiation	Automation
Connect	Magic	Diversification	Open innovation
Learn	Delivery		Security

The Takeaway Box: Technology Is a Game Changer; Get It to Work for You Before Your Competition Does

- Technology is used to support business models. It also works the other way—several innovations are made possible and built around technology, for example, location-based services are built around mobile technology and gaming/simulation is based on AR/VR. The two-way relationship between business landscape and technology is the major driver for future business growth.
- Develop a very good understanding of the technology and enterprise landscape. Especially, keep a keen watch on the key technologies and on every sign of how the competition may be planning to use them.
- Get a good grasp over what are the strategies you need to build to use the technologies in your business, the organization to build, talents to be acquired/hired, the training needed, and the costs and timelines.

References

AI Branches. Retrieved from https://en.wikipedia.org/wiki/Outline_of_artificial_intelligence#Branches_of_artificial_intelligence.

AI Index. (2016). Retrieved from https://aiindex.org/.

AI100. (2016). One Hundred Year Study on Artificial Intelligence. Retrieved from https://ai100.stanford.edu/.

Alternatives to Internet. Retrieved from https://github.com/redecentralize/alternative-internet.

AR Brief. Retrieved from https://en.wikipedia.org/wiki/Augmented_reality.

Birudavolu, S. (2015). Open Innovation in ICT: An Empirical Assessment of Global Telecommunications Services. Unpublished thesis submitted for PhD at Indian Institute of Foreign Trade, New Delhi.

Cai, Y., & Goei, S. L. (2012). *Simulations, Serious Games and Their Applications.* Springer, Gaming Media and Social Effects, ISBN 978-981-4560-32-0.

Cai, Y., Goei, S. L., & Trooster, W. (Eds.). (2017). *Simulation and Serious Games for Education.* Springer, ISBN 978-981-10-0861-0.

CIO Magazine. Retrieved from http://searchcio.techtarget.com/magazineContent/How-Company-Size-Relates-to-IT-Spending.

Cloud. Retrieved from https://en.wikipedia.org/wiki/Cloud_computing.

Containers. Retrieved from https://www.cio.com/article/2924995/software/what-are-containers-and-why-do-you-need-them.html.

Devbattles, An Absolute Beginner's Guide to Machine Learning, Deep Learning, and AI. Retrieved from https://www.devbattles.com/en/sand/post-3948-An_absolute_beginners_guide_to_machine_learning_deep_learning_and_AI.

DoT (Department of Telecommunications). National Telecom M2M Roadmap. (2015, May). Ministry of Electronics and Information Technology. Retrieved from www.dot.gov.in/sites/default/files/National%20Telecom%20M2M%20Roadmap.pdf.

Druzhinin, A. (2017, September 1). Whoever Leads in AI Will Rule the World. RT website. Retrieved June 30, 2018, from https://www.rt.com/news/401731-ai-rule-world-putin/.

Gartner. (2017). IT Budget Sample. Retrieved from http://www.gartner.com/downloads/public/explore/metricsAndTools/ITBudget_Sample_2012.pdf.

Gartner. (2018). Benchmark Analytics industry standard IT Metrics. Retrieved from https://www.gartner.com/doc/3832769/it-key-metrics-data; https://www.gartner.com/doc/3832474/it-key-metrics-data.

Genes in Space. Cancer Research UK. n.d. Retrieved from http://scienceblog.cancerresearchuk.org/2014/02/04/download-our-revolutionary-mobile-game-to-help-speed-up-cancer-research/.

Georgia AI. (2017). *AI Recreates Video Games, Georgia Institute of Technology.* Retrieved from https://www.theverge.com/2017/9/10/16276528/ai-video-games-game-engine.

Ghoshal, A. (2018, January 19). Microsoft's Artificial Intelligence Application Can Now Turn Text into Images. Retrieved from https://www.vccircle.com/need-for-laws-to-regulate-advances-in-artificial-intelligence-microsoft/.

Good, B. M., Loguercio, S., Griffith, O. L., Nanis, M., & Su, A. I. (2014, July 29). The Cure: Design and Evaluation of a Crowdsourcing Game for Gene Selection for Breast Cancer Survival Prediction. *JMIR Serious Games, 2*(2), e7. https://doi.org/10.2196/games.3350.

Google Cat Videos Recognition. (2012). Retrieved from https://www.wired.com/2012/06/google-x-neural-network/.

Greenspan, G. (2015, November 22). Avoiding the Pointless Blockchain Project. Retrieved June 25, 2018, from https://www.multichain.com/blog/2015/11/avoiding-pointless-blockchain-project/.

HIT. (2018). Human Interface Technologies Team, University of Birmingham, Serious Games at the University of Birmingham. Retrieved from www.birmingham.ac.uk/Documents/college-eps/eece/research/SeriousGamesattheUniversityofBirmingham.pdf.

Hsu, Chin-Lung and Lin, Judy Chuan-Chuan. (2016, September). An Empirical Examination of Consumer Adoption of Internet of Things Services: Network Externalities and Concern for Information Privacy Perspectives. *Computers in Human Behaviour,* Elsevier, 62, 516–527.

Huffington Post. (2017, October 19). *How AI is Changing the Way We Do Business.* Retrieved from https://www.huffingtonpost.com/entry/how-artificial-intelligence-is-changing-the-way-companies_us_59e8d1f2e4b077c789918b4d.

Hutson, M. (2017, September 7). The Future of AI Depends on a Huge Workforce of Human Teachers. *Bloomberg,* https://www.bloomberg.com/news/articles/2017-09-07/the-future-of-ai-depends-on-a-huge-workforce-of-human-teachers.

IANS. (2018, January 23). *Artificial Intelligence Can Boost Revenues by 38%, Employment by 10%: Accenture.* Retrieved from https://economictimes.indiatimes.com/tech/ites/artificial-intelligence-can-boost-revenues-by-38-employment-by-10-accenture/articleshow/62622147.cms.

IBM Blockchain, Blockchain Explained. Retrieved from https://www.ibm.com/blogs/blockchain/category/blockchain-education/blockchain-explained/.

Immersive Technology Wikipedia. Retrieved from https://en.wikipedia.org/wiki/Immersive_technology.

Internet's Fundamental Limitations. (2010). EC FIArch Group. Retrieved from www.future-internet.eu/uploads/media/FIArch_Current_Internet_Limitations_V0_9.pdf.

IoT Protocols. Retrieved from https://www.postscapes.com/internet-of-things-protocols/.

IPFS Resources. (n.d.). Retrieved from https://blog.neocities.org/blog/2015/09/08/its-time-for-the-distributed-web.html; https://www.sitepoint.com/http-vs-ipfs-is-peer-to-peer-sharing-the-future-of-the-web/; https://ipfs.io/.

IPv6 Migration Issues. Retrieved from https://www.6connect.com/resources/top-5-concerns-of-network-admins-about-migrating-to-ipv6/.

IPv6 Security Issues. Retrieved from http://ipv6now.com.au/primers/IPv6SecurityIssues.php.

Jones, K. (2017, June). GANGogh: Creating Arts with GANs. Retrieved from https://towardsdatascience.com/gangogh-creating-art-with-gans-8d087d8f74a1.

Joyce, K. (2017, June 5). *AR, VR, MR, RR, XR: A Glossary to the Acronyms of the Future.* Retrieved from www.vrfocus.com/2017/05/ar-vr-mr-rr-xr-a-glossary-to-the-acronyms-of-the-future/.

Karasavvas, T. (2017, July). Level Up: How Video Games Evolved to Solve Significant Scientific Problems. Retrieved from Arstechnica.com; https://arstechnica.com/gaming/2017/07/level-up-how-video-games-evolved-to-solve-significant-scientific-problems/.

Kura Eclipse Framework. Retrieved from http://eclipse.github.io/kura/intro/intro.html.

Lang, A. S., & Bradley, J.-C. (2009, October). Chemistry in Second Life. *Chemistry Central Journal, 3,* 14. https://doi.org/10.1186/1752-153X-3-14.

Linden Labs. *Second Life.* Retrieved from www.secondlife.com.

Meyer, D. (2017, September 4). Putin Says AI Leaders Will Rule the World. *Fortune*. Retrieved from http://fortune.com/2017/09/04/ai-artificial-intelligence-putin-rule-world/.

Microsoft AI. (2018). Retrieved from https://www.bloomberg.com/news/articles/2018-01-18/microsoft-says-ai-advances-will-require-new-laws-regulations.

Microsoft Blog on Telecom Operators. Retrieved from https://blogs.microsoft.com/firehose/2013/05/13/microsoft-helps-telecom-companies-find-new-revenue-streams-in-the-cloud/.

Miglani, S. (2018, February 18). Indian Banks Could Take A Massive $3 Billion Hit From Punjab National Bank Fraud. *Huffington Post*.

Money Control, Fraud at PNB May Cost Banking Sector Much More Than Rs 11,400cr. (2018, February 24). Retrieved from http://www.moneycontrol.com/news/india/fraud-at-pnb-may-cost-banking-sector-much-more-than-rs-11400cr-2515589.html.

P2P—Peer-to-Peer. Retrieved from https://en.wikipedia.org/wiki/Peer-to-peer.

Plansky, J., O'Donnell, T., & Richards, K. (2016, January 11). *The Strategist's Guide to Blockchain*. Retrieved from https://www.strategy-business.com/article/A-Strategists-Guide-to-Blockchain?gko=0d586.

Popper, N. (2016, June 26). *How China Took Center Stage in Bitcoin's Civil War*. Retrieved from https://www.nytimes.com/2016/07/03/business/dealbook/bitcoin-china.html.

Robotics. Retrieved from https://en.wikipedia.org/wiki/Robotics.

Senevirathne, G. S., Kodagoda, M., Kadle, V., Haake, S., Senior, T., & Heller, B. (2011, January 27). Application of Serious Games to, Sport, Health and Exercise. *Proceedings of the 6th SLIIT Research Symposium, Sri Lanka*. Sheffield Hallam University, Sheffield (United Kingdom). Retrieved from http://shura.shu.ac.uk/3672/1/Application_of_serious_gaming_to_Sport_Health_and_Exercise.pdf.

SGIL. (2018). Serious Games International Limited. Retrieved from www.seriousgamesinternational.com.

Smart Contracts and the DAO Implosion. (2016). Retrieved from https://www.multichain.com/blog/2016/06/smart-contracts-the-dao-implosion/.

SpiceWorks. (2017). Retrieved from www.spiceworks.com/marketing/state-of-it/report/.

Statista AI. (2016). *Revenues from the Artificial Intelligence (AI) Market Worldwide, from 2016 to 2025 (in Million U.S. Dollars)*. Retrieved from https://www.statista.com/statistics/607716/worldwide-artificial-intelligence-market-revenues/.

Stinchcombe, K. (2017, December 22). *Ten Years In, Nobody Has Come Up with a Use Case for Blockchain*. Retrieved from https://hackernoon.com/ten-years-in-nobody-has-come-up-with-a-use-case-for-blockchain-ee98c180100.

TechRepublic, CIO/CTO definition. Retrieved from https://www.techrepublic.com/blog/tech-sanity-check/sanity-check-whats-the-difference-between-cio-and-cto/.

The Cure, Gene Games. Retrieved from http://genegames.org/cure.

TMForum Frameworx. Retrieved from www.tmforum.org/tm-forum-frameworx-2/.

TOGAF 9.1. (2011). *The Open Group Enterprise Architecture Framework*. Retrieved from http://www.opengroup.org/subjectareas/enterprise/togaf.

Upson, S. (2016, December 22). *The AI Takeover is Coming. Let's Embrace It*. Retrieved from https://www.wired.com/2016/12/the-ai-takeover-is-coming-lets-embrace-it/.

Urban, T. (2015). The AI Revolution: The Road to Superintelligence. Retrieved from https://waitbutwhy.com/2015/01/artificial-intelligence-revolution-1.html.

VR Brief. Retrieved from https://en.wikipedia.org/wiki/Virtual_reality.

Whitehouse. (2016, December). Office of the President of the USA. Artificial Intelligence, Automation and the Economy. Retrieved from https://www.whitehouse.gov/sites/whitehouse.gov/files/images/EMBARGOED%20AI%20Economy%20Report.pdf.

Wilson, J., Daugherty, P., & Morini-Bianzino, N. (2017, March 23). The Jobs That Artificial Intelligence Will Create. *MIT Sloan Review*.

Wireless Community Network. Retrieved from https://en.wikipedia.org/wiki/Wireless_community_network.

Wireless Community Network Implementations. Retrieved from https://en.wikipedia.org/wiki/List_of_wireless_community_networks_by_region.

Zago, M. G. (2018, January). Why Web 3.0 Matters and You Should Know About It, Medium. Retrieved from https://medium.com/@matteozago/why-the-web-3-0-matters-and-you-should-know-about-it-a5851d63c949.

## CHAPTER 5

Finding Business Focus with ICT

A designer knows he has achieved perfection not when there is nothing
left to add, but when there is nothing left to take away.
—Antoine de Saint-Exupery

5.1 Introduction

In today's fast-paced world, only businesses with extreme focus succeed (Ries 1996). Very few companies with a "general" focus can make it. The exceptions are century-old corporations, like GE or Tata, which have captured many opportunities (especially government contracts) and established a lead through several decades (say from 1890 to 1960) when no serious competitor was around then. The regulations at the time also permitted such monopolies in the interest of the nation's development, in contrast to the current day policies that strongly encourage competition in the market. There are also companies that pursue a limited form of "generality", but they are first movers in the market like Amazon, which is a company that continues to sink in enormous money into technology and has only just become profitable (in 2016) having been founded over two decades ago in 1994. Amazon has established itself strongly for the future through a heavy ICT base. In fact, ICT has enabled innumerable new business models and is disrupting the market beyond recognition.

Far likely to succeed in the current day and age are companies that target specific, even niche markets and become very strong in it. The long tail

© The Author(s) 2019
S. Birudavolu, B. Nag, *Business Innovation and ICT Strategies*,
https://Doi.org/10.1007/978-981-13-1675-3_5

of the market that caters to an infinite number and variety of increasingly narrower segments of the market, seems to be more profitable. The narrowest market segment possible is that which is reduced to one person, meaning that the product or service is completely personalized and customized to the individual level. This would, in principle, provide the ultimate customer satisfaction.

The focus area may have several applications in the market that one did not imagine in the beginning of the journey. It is important for an organization to keep its eyes and ears open and scan the market for possible new and different use-cases for its product.

For example, products built for defense purposes are now used in a variety of civilian applications. Drones and remote communication systems are now being re-used for civilian security.

The USA's NASA (National Aeronautics and Space Administration) is required to publish its significant research findings to the public, because public tax money goes into the NASA's research programs. From the past three decades, the NASA publishes them in the form of Tech Briefs (NASA Tech Briefs 2018). The industry eagerly laps up the research to find ways to apply them in different fields. These spinoffs, tech transfer resources, and application stories are immensely popular, and many have yielded success.

How can organizations find focus using ICT and open innovation? This is the subject of this chapter.

5.2   Business Focus: Summary of the Process

The business focus is based on three strategic aspects: focused thinking, decision-making, and transformation in the leadership. Table 5.1 provides a snapshot. While considering ICT as a major game changer in business operation, companies need to think strategically how they can use ICT as part of their core strategic approach. In all such cases, firms need to go beyond the traditional thinking process, but without losing focus. Similarly, while using ICT tools for decision-making, firms are required to open up with respect to various opportunities unfolded by ICT. Firms need to judge these opportunities and then take a logical decision how to move ahead. In this context, changing pattern of leadership is crucial. Managers need to be prepared and encouraged constantly for the change. They need to be nudged towards more imaginative business models and appropriate reward structures must be setup. A major structural change in approach with a clear focus can help an organization to be ready with twenty-first-century business reality. The following table describes the major points in this regard.

Table 5.1   The process of building the business focus

Driver for focus	Framework	Process for business focus
Focused thinking	Fact or intuition based	• Will it be only with "bounded reality", that is, thinking based on rational constraints the organization is facing? • Will it be more intuitive and innovative? • Engaging management in the thinking process
Focus on decision-making	Self-interest, chaos, or structure	• Problem and opportunity recognition • Analysis of cause • Identification and evaluation of solutions (with or without ICT tools) • Identifying optimal solution and implementation of the same
Focus on leadership dynamics	Managers vs. leadership in delivering the strategy	• Exploring key capabilities and strength of managers and identify how to adopt new approaches and avenues • Encourage stability, handle complexity, and promote innovation • Help in improving performance of managers and propose useful changes including in ICT strategies • Focus on internal and external alignment • Promote rational thinkers, motivate, and provide rewards

Source: Adapted and modified from *Strategy and Strategists* by James Cunningham and Brian Harney, Oxford University Press, 2012

5.3   How Companies Lose Focus

Organizations in every sector, and even people and governments today, need to be strongly focused. Companies that don't do this may be well on their way out. The churn in the list of Fortune 500 companies in the past two decades is a stark indicator.

Unfortunately, even successful tech companies today are steadily losing focus, if you go by the number of their market offerings. This includes companies like Microsoft, Google, Facebook, Apple, and so on, and these may face the same fate of erstwhile tech giants if they're not careful. One only needs turn the pages of the past decades to look at the "invincible" companies of yesteryears, at a time when Google and Facebook were start-ups with little cash at the time when these companies were behemoths with billions of dollars at their disposal. And companies like Uber, Alibaba, and Tesla were non-existent.

Many of the old, large corporations in several sectors are under a tight squeeze owing to relentless pressure on market share/margin/revenues/ market value for years now. This pattern cuts across sectors—retail, Telecom, airlines, banking, energy, high tech, automotive, and so on.

Examples:

- *Retail*: Marks & Spencer, Woolworth, Macy's, Aeon, and so on
- *Airlines*: Air India, Delta, Emirates, Alitalia, Qatar, United Airlines, and so on
- *Banking*: Bank of Japan, State Bank of India, Brazilian Banks, Eurozone Banks, Banks in USA (where five US banks have already failed by July 2017)
- *Telecom*: AirTel, BSNL, Verizon, Ericsson, NSN, Vodafone, Deutsche Telekom
- *High Tech*: IBM, Hewlett-Packard, Dell, Samsung, Toshiba
- *Energy*: ExxonMobil, PetroChina, Shell, General Electric
- *Automotive*: General Motors, Toyota, Hyundai
- *Hospitality*: Hilton Hotels, Shangri La Hotels, Hyatt Hotels, InterContinental Hotels Group

Many have been severely disrupted by ICT, and also many have lost focus.

Some have successfully transformed themselves with ICT and innovation and have not merely survived but are thriving well. As mentioned in the introduction, the trade-off between technological capabilities and market capabilities/knowledge are the main drivers for success in such cases. Companies may still be able to sail through with slower innovation pace if they have better control on the market.

Many of the upstarts who have now become heavyweights are at risk of going down the same path. History repeats itself, if the lessons are not learned. With success, the existing models and mindsets get entrenched, and even complacence set in. The money and market power seem to accumulate automatically! One feels whatever one touches will become gold, owing to one's innate greatness (and brand power)! Then the board hires top executives and sets quarter-to-quarter targets. The easiest way to achieve targets is through product/service line extensions, expansion into non-core areas, rampant mergers and acquisitions, that is, defocus! The company loses power, and market share, and thereby paves its own downfall.

However, there are large and highly profitable companies that have managed to retain a deep focus over the decades.

Examples of large, profitable companies with deep focus are:

- Bose—the maker of premium speakers. Revenue US$3.8 billion
- Netflix—streaming media/video-on-demand. Revenue: US$11.69 billion (2017)
- Berkshire Hathaway—investment company. Revenue: US$223 billion (2016)
- The Lego Group—toys. Revenue US$2.1 Billion (2017)
- Nvidia—GPU (graphical processing unit) chips. Revenue: US$6.91 billion (2017)
- Southwest Airlines—low-cost airlines. Revenue: US$21 billion (2017)
- Alibaba—e-commerce. Revenue US$26.42 billion (2017)

Focus works. In fact, Apple really took off only after Steve Jobs rejoined the company and cut down the number of product offerings offered to merely six products. In other words, he increased focus. Constantly cutting prices to deal with competition doesn't help in the long run, and will cast one into a downward spiral as many online retail giants in India have discovered. The only exit possible for such companies is to get sold out to a bigger player.

5.4 FRAMING THE FOCUS PROBLEM

With increasing complexity of business, depth of domain know-how needed, and the threat of severe competition, it is futile for any organization to expect to stay ahead in the game unless one keeps at it every day. Hence no organization can afford a dilution of focus.

Even to remain in the game, one must have a strong, narrow deadly focus, not an approach of "all things to all people, anytime, anywhere". Develop focus around one expertise. Think www.leftyslefthanded.com (not Walmart) in order to keep up healthy profitability. Leftylefthanded. com supplies custom goods for left-handed people, for example, scissors and tools. Only a very few players like Walmart or Big Bazaar can achieve profits through low margins and sheer scale. And that's a *very, very* big, difficult game.

In fact, you will see that focus is happening all around you, from Coffee Day to Cream Stone ice creams to OYO Homes. Focus is an entry-criterion now. It's not the end-game, but generally companies with focus prevail. Indian companies, protected by a thousand government regulations in the past decades, have a strategy of offering everything from "Insurance to Jewels to Textiles to Mobile Services". They're all poised to get eaten alive by much leaner and meaner competitors with focus (e.g. CaratLane, Bluestone). Barring exceptions, it is difficult to find many products with Indian origin (brand name) in foreign markets. Whereas you'll easily recognize and pick up a Cadbury or a Dell, precisely because they're focused. Bose Speakers will not enter into construction business because "there is money in it". If you were to look at the latest market share of car companies in India and their profitability, you'd get the idea.

ICT enables you to do both—have a narrow focus and also address a larger market. A hypothetical example is to sell specialized tools for left-handed people all over India (or even beyond) through E-Commerce. One can identify more easily with the customers, understand them better, and build content that targets them directly. The more the narrow and specific the topic is, the easier it is to talk to the customer about his/her user journeys and needs, address pain points, and solve problems. It's easier to attach value also to more concrete, tangible things.

Some real-life examples:

- www.16stitches.com is an online retailer that sells custom made shirts.
- www.Lenskart.com specializes in eyeglasses and eyewear.
- www.paytm.in is a Paytm mobile app—monetary, cashless transactions using the mobile phone.
- www.gempundit.com—gemstones online.
- www.SaffronArt.com—online market place for artwork.

With open innovation, you can set up collaborations, on all sides—suppliers, partners, customer networks, re-sellers, and so on (Birudavolu 2015). To continue the hypothetical example, if you're good at making and selling goods for left-handed people, then you can also tie up with Amazon, eBay, and so on, to expand your market. You could build a mobile app which interested users can download. Then do a tie-up with AirTel and other Telcos to target their subscriber base (for a fee). One can tie up with channels for Digital Marketing. You'd find a decent, profitable, and sustainable customer base. But how do we go about doing it? This chapter addresses this question.

5.5   THE PROCESS FOR FIRST DEVELOPING A FOCUS

It is simple to define a process for developing a focus. The problem is that organizations do not adhere to the basic steps, due to the other reasons discussed in this chapter, such as distractions, problem of too much choice, temptation for product line extensions to meet immediate goals, and so on. There is no real market study done to establish the need first. Interestingly, the top reason why startups fail is because there is no market need (Lance 2016).

The entire process and all the steps within it are iterative. The iterations should continue until the product clicks in the market, and then one should continue to refine it anyway. ICT is of great help throughout the process.

5.5.1   Establish the Strong Market Need (Market Research)

The market you're targeting should be centered around a *strong unmet need*. This could typically be a critical customer pain point. The spectrum of needs ranges from:

1. *High Criticality*: Critical for survival or very strong need (i.e. cannot do without). Think life-drugs, basic food/clothing/shelter/transport, hospital emergency, ambulance. Not meeting these needs will very severely hamper the customer's life and career. It's important to note that one customer's survival needs could be another's nice-to-have needs. For example, Wi-Fi connectivity could be a luxury for the bottom-of-the-pyramid people but a basic need for a business that needs to meet business deadlines daily.
2. *Medium Criticality*: Good need. These are those, without which, the customer will get affected but can still carry on with his life, despite the burden or pain. Over time, the pain may get to a critical point, where he'll probably decide to find a solution. Many needs start out in this category, and progress to the critical stage, as solutions become available and affordable.

3. *Low Criticality.* Nice-to-have. These are extras or luxury goods. They come into picture in a customer's mind after his more critical needs have been met. Affordability, perception, and other soft factors come into play.

Criticality can also arise suddenly, such as when a regulation changes, then software needs to be updated to reflect the change in regulation immediately. For example, now all tax and accounting software in India need to cater to GST, and the software vendors need to upgrade the software within a stipulated deadline.

Thus, look for:

- Unsolved problems that are afflicting the customer
- Gaps (unaddressed areas) in the current market offerings (solutions), by functionality, price, performance, access, customer profile/segment, and so on

The gap could also be a niche in the market, such as that occurring in the long tail of the market, for example.

Define the problem statement clearly and identify the customers it is targeting. To do this one needs to talk to as many of the potential customers as possible and take notes to extract the essence of the problem. Most of the time, the difficulty is that one has not identified the real problem (customer need) correctly. As the saying goes, the customer does not need a quarter-inch drill, he needs a quarter-inch hole. The drill is only a means of getting him the hole.

It is important to note that the solution space and the problem space are very different. Without defining the problem properly, one should not embark on the solution. The reason for this is that one may discover an extremely simple solution if the problem is thoroughly understood.

The Problem Space has terms like pain, gap, problem, need, urgency, priority, demand, criticality, struggle, cumbersome, time-taking, slow, expensive, cost, difficult, changing, overwhelming, primitive, basic, fragmented, suffering.

The Solution Space has terms like idea, solution, innovation, product, technology, re-use, offering, automation, integration, seamless, comfort, low cost, inexpensive, effortless.

The Terms Common to the two spaces are systems, workflow, return-on-investment, business, end-customer, definition, market, customer experience.

The terms that are common to the two groups are typically used to compare the two spaces before and after deploying the solution, for example, ROI or customer experience before and after the solution. Obviously, the comparison should commence right from the time of choosing the solution, not after embarking on the solution or deploying it.

5.5.2   Identify the Customers

It is essential to zero down on the most likely set of customers who will have a dire need for this product. Define this set in as many ways as possible. Identify attributes, for example, by age, gender, educational qualifications, geography, and so on. Find out the actors and factors in the ecosystem that influence the customers, and the ways they influence them. This could be their jobs, friends, relatives, institutes, transportation they use, kind of mobile phones, the apps/software/websites they use, and so on. Constantly extract patterns. Every bit of information about the customers is valuable. Gain insights. Constantly refine your understanding of the customers.

5.5.3   Find Channels to Reach the Customers (Channel Strategy)

The product needs to reach the different segments of potential customers. Each segment may need a different customer experience. For example, mobile users and PC users. And again, smartphone users and those using basic mobile phones. Once you've identified the customers thoroughly as given in the previous step, this should be easier. Develop a strong channel strategy. One of the thrusts of this book is open innovation. Collaborate and co-create with your partners to reach customers.

5.5.4   Find a Solution

It is good to use design thinking to find a solution. The process is briefly explored in another chapter, titled "Winning the Competition". There are many good techniques for ideation too. The solution should have a good product-market fit. For accomplishing this, one must first have a value hypothesis, that is, ask the questions "what is the product idea?", "who does it address?", and "how does it address it?" Then a survey or preferably a prototype can be built to test the hypothesis in the market. Their feedback is of paramount importance. The key question to test customer delight is whether they would sell or refer the product to others. It is important to find effective metrics for the product. The metrics for a website will be different from that of a mobile app, which will be different for a hardware product. The most important ones are customer-centric metrics, such as customer's time or money or effort saved, or revenue earned for him. The metrics should obviously be designed to measure a product-market fit. It is best to strive and get an outrageously good product-market fit, rather than chase being first to market or something else (Griffin 2017).

It should be clear what one is launching or taking to the market. Is the launch intended to test the product validity in the market, that is, MVP (minimum viable product) or is it to showcase the robustness of the product? The results of the launch should be used to refine the product and align with the customers even better.

It is always good to choose a larger pool to fish in. Giant markets are great. Prefer Blue Oceans (Kim and Mauborgne 2005). That is, choose markets with low competition and vast potential as opposed to highly competitive markets.

5.5.5   *Important to Innovate to Find a Unique Solution*

A unique solution is:

- What the customer wants,
- Which is currently unavailable in the market (no comparable offerings from the competition), and
- What your product does well.

Differentiation in the market is key. The other chapters in the book explore ways to innovate and find unique solutions, including open innovation.

5.5.6   *Is Your Product Viable?*

The game must be worth the candle. Many companies run into financial trouble even after years of scaling. In the world of startups, there are many whose products won't be financially viable. On the cost side, the unit economics and the cost of selling must work out when compared to the revenue side, that is, pricing and scale. Financial projections are at the heart of the business model. This is especially true of startups, where it is a question of survival of the business and they cannot afford to make many mistakes. The Traction Gap Framework (Wildcat Ventures 2017) is a very useful framework that helps startups create the Minimum Viable Traction. This framework is described briefly in the chapter on strategy.

5.5.7   *Is Your Product Competitive?*

There is an entire chapter in this book on how to win the competition in the marketplace. Briefly, the key things to consider for making your product competitive are on how the customer meets his needs currently (in the absence of your product), and the alternatives available to him. Every single one of these ways constitutes competition to your product, regardless of whether they are low-tech, manual, or entirely different from your product.

For example, the competition to a high-tech software that uses AI/ML could be a simple paper-based system, or a small mobile app that is free for download, or an old legacy system with a simple UI. It is likely that another customer segment may never need the software at all because they are happy phoning a customer contact center and talking in their local language to get their information.

5.6   Business Model Canvas

The Business Model Canvas or BMC (Osterwalder and Pigneur 2010) is an excellent way of documenting, discussing, and refining the business model. It is a visual map describing nine elements (or building blocks), in three categories:

- *Value and Customers*: four blocks—a firm or product's value propositions, customer segments, customer relationships, channels
- *Scope and Capability*: three blocks—key activities, key resources, key partners
- *Economics*: two blocks—cost structure and revenue streams

BMC should be used to refine focus and understand your business constantly. It is good to carry out regular BMC exercises with the teams. For startups, a scaled down version, Lean Canvas Model (Maurya 2018) may be suitable, for example, they may not have partners in the early stages. The Lean Canvas also has nine elements:

- Unfair advantage, customer segments, channels
- The problem, solution, key metrics
- Cost structure, revenue streams

5.7   Preventing and Removing Inessentials

This is the concept of encashing and concentrating existing comparative advantage and discovery of new advantage. To build and retain a sharp focus, the following must be considered enemies and should be dispensed with immediately:

Redundancies, features that are unimportant/unused/little-used/obsolescent/fringe/good-to-have features, trivia, wastage, meaningless complications.

Removing these is a much harder problem to have than avoiding a buildup of superfluousness in the first place. It's essential to put systems and processes in place to filter out not only noise but also well-meaning solutions that make temporary sense. In companies that build products or offer services, a strong product management function helps. The term product here includes service offerings too, that is, a service bundle designed, crafted, and offered as a commercial service pack.

This prevention/selection of features must be driven from the top, especially in large companies, for chiefly two reasons:

1. So that the middle management does not buckle to pressure from the customers and pile on features into the product
2. So that the middle management does not engage in product line extensions in order to meet quarterly sales targets

For the first one, the product manager must keep the Product Roadmap simple, and strictly allow only those features that:

- Are strongly aligned to the focus of the organization
- Are needed by most customers and stakeholders

For the second one, the executive management must rule with a strong hand, and disallow all product extensions in general. The few really important ones must go through painstaking review and explicit approval. These must be very far and few.

Removing inessentials from an existing implementation/process/offering/product/service is fraught with risks and dangers of every kind—technical, political, financial, contractual, and so on. One would run up against entrenched behaviors and habits, breaking the existing implementation, investments to redo and test, impact on marketing and communication, and even HR practices if relevant teams need to be retrenched or relocated. There could be a few entrenched customers who are using that feature since years and will not let go off so easily.

Overall, the Product Roadmap must be communicated well, and expectations managed. It's good to use tools like Pareto Analysis (80/20 Rule), Dependency Matrix, UI/UX tools and Analytics to take product management decisions.

Features that are being sunset should be clearly informed to all, well in advance, especially to the subset of the customer base that is using the features. They will need help for migrating their deployment to the new roadmap. They will also need to budget for consultancy, redesign, migration, and testing.

Despite all these challenges, an organization should not shy away from re-factoring and simplifying their product lines and roadmaps. Constant change is a fact of life, and much scope creep and technical as well as business obsolescence is bound to happen over time. There could be changes at various level in the organization itself. To keep the central focus intact, the steps described must be taken, albeit cautiously.

5.8   PIVOTING, AND HOW IT IS DIFFERENT
FROM DEFOCUSING

Occasionally when a company's business model is no longer working as expected, it may need to pivot, that is, change its focus, its direction, in substantive ways. This means that the company has decided to pursue a new business model. As per Steve Blank:

> *A pivot may mean you changed your customer segment, your channel, revenue model/pricing, resources, activities, costs, partners, customer acquisition—lots of other things than just the product.*

Pivoting means that the organization has found a new focus that works better; it does not mean that it has become fuzzier in its focus or that it has lost focus. It is not about product line extension or using one brand name to cover multiple, disparate product lines. Pivoting is usually a drastic change in course. It results in a near complete transformation of the business.

In the Lean Startup Model (Ries 2011), there are four prescribed steps (with iterations) for building a startup: Plan-Build-Measure-Learn. This is followed by another step with two options: Persevere or Pivot. If the first four steps go as planned and the results are good, then the startup may proceed with the idea, that is, Persevere. Whereas if the iterations show that the hypothesis is not working out, then the idea may need refinement, that is, startup needs to Pivot.

While it is easy to understand pivoting in the context of a startup or a small/mid-sized organization, one needs to be very careful while pivoting a large organization. A startup, by definition, is exploring and testing its business model, and hence has hypotheses that are not fully formed. Thus, it starts off with one model, and pivots to adjust the model to the market. On the other hand, a large company may pivot when its hitherto successful business models are reaching maturity and are on the verge of declining.

Some examples of pivoting in large companies are:

1. IBM decided to get out of its declining PC business and concentrate on enterprise products/services. This is an example of improving focus. It has paid off well for IBM.
2. Intel moved out of memory chips and focused on microprocessors. It was a hugely successful step.
3. Starbucks Coffee started off by selling coffee beans and expresso coffee makers. They completely transformed themselves into a premium coffee-house chain.
4. Wrigley moved from being a seller of baking powder and soap into a chewing gum-only business.
5. Instagram was a check-in app that had features of gaming and photos, but they decided to focus on photo-sharing only.

If large companies decide to pursue new markets with new product lines, it is far better to launch a different company to focus on the new product. This can happen if they find, for their products/ideas, entirely new or unexpected applications in a different field. It is not at all a good idea to use its existing brand name and engage in product line extensions. That would result in defocusing the company, and is, in general, not a sound strategy. Another option is to collaborate with another company that is in a different line, thereby retaining one's own focus. Sometimes it is likely that the company has hit upon one goldmine of a market for its application. It may decide to pursue only that avenue as it is far more valuable than the others for a significant time to come. Chasing the other less lucrative avenues may dilute its efforts and focus. It is possible that the general market is getting commoditized, and hence it needs to specialize fast. For example, a manufacturer of drones may find that its camera/video drones have much appeal in the market. So, it may choose to abandon general purpose drones and focus on perfecting its camera/video drones. Under no circumstances should a company fall into error of an "anything, anytime, anywhere for anyone" type of thinking. The competition is likely to eat it alive.

5.9   Connectivity Can Increase Focus

This section deals with the difficult subject of:

• How to keep things simple and the focus intact
• While exploiting the abundant opportunities in the market
• By leveraging ICT and open innovation
• And unleashing innovation within the organization
• To make rich and relevant offerings in the market

Contrary to popular misconception that ubiquitous connectivity diffuses concentration, it can actually *increase* focus. The misconception is fueled greatly by the abundant distractions provided by social media, entertainment/news/shopping websites, and other time-wasters that play havoc with attention and bandwidth. And this translates to marketing and product management too. There is a problem of plenty when choosing which features to include in the product. The market has a plethora of offerings in each area, even if the other offerings are not in the same category as our product to compare with. For example, there are a variety of ways to build the UI/UX, choose underlying platforms and operating systems. For instance, in the latter, one can choose proprietary (Microsoft, IBM, Solaris, etc.) or open source (Red Hat Linux, Ubuntu, Debian, etc.) And again, these are constantly evolving.

A surfeit of choice can indeed shatter focus and increase waste, especially human resources. For example, instead using ten vendors available in town, one can evaluate a thousand online vendors offering attractive deals. It takes effort and time to process offers, negotiate, and close a deal. And shortlisting a vendor may not even be the core part of the business to warrant such wastage of bandwidth. Sometimes a multitude of choices can result in poor decisions. For instance, if offered 50 different cars to choose from, one may get tempted to attach greater weightage to look and choose a pricier option than if one had a choice of only five cars that happen to fulfill all the requirements.

However, the above problems generally arise due to two reasons:

- Lack of clarity
- The entire process is not managed well

Both, if tackled well, can open up a treasure trove of opportunities. And Information Technology provides ample means to solve both.

The first step for achieving clarity is to acquire a high degree of self-knowledge:

A thorough understanding of ones' origins, background, training, size, capabilities, assets, liabilities, history, successes, failures, alliances, market positioning, competitive analysis, revenues, market share, penetration, geographical reach, regulatory and standards compliance, access to resources, and so on.

Tools like SWOT analysis and TOGAF's Capability Model really help in defining oneself. With this, the core of one's self begins to get identified. It is good to start at one extreme and define what one can do that others can't do. The whole idea is to identify the as-is situation (status quo), define the to-be situation (desired state), and do process engineering to achieve that.

The second step is to identify the possibilities based on one's core capabilities and the markets. Typically, one asks several what-if questions, such as what if we increased our reach in different ways through partnerships, expansion, improving products, innovation, and so on.

The key is to remember that the focus needs to be as specific and as narrow as possible. Expansions and improvements here do not refer to product line extensions or getting defocused and randomly stepping into areas outside one's core focus. It refers to using one's focus in different ways to build new and innovative business models and target different markets.

All this results in further refinement in clarity, much as when a diamond is polished and shaped from different directions, it results in greater lucidity and brilliance. It's important to make this a constant process and not a one-time activity. Many a time there are several paths to choose from. There is a need to experiment, prototype, and test.

Often, the ability to form and use links, build an ecosystem, and provide unique, superior experience is a *core* capability by itself! And the product or service that one has is only a tool to drive the core capability. For example, a maker of medical instruments may realize that the core capability is to provide superior diagnostic and treatment experience, rather than merely selling instruments. Thus, they can build the whole ecosystem around the set of medical devices, sometimes even partnering with the makers of other medical devices.

The Internet offers practically unlimited possibilities for collaborations, extensions, delving into different areas, and extending/morphing business models. Until about a couple of decades ago, before the web had caught on, *getting* information was the major problem. With widespread connectivity and plenitude of web content available, the problem has shifted to that of *selection* of information. Many a time, having too much information is as good as having no information. A search for the keyword *innovation* on google will show links to over 649 million websites in 0.53 seconds.

It is easy to get defocused.

The key then is to use tools to *increase* focus.

Filters, checklists, frameworks, models, standards, regulations, trends, technologies, emerging forces, and so on all serve to help in constraining the universe of possibilities to a manageable set, thereby increasing focus.

For example, there will only be a handful of technologies that will suit one's needs depending on a known list of factors such as price, affordability, availability of experienced manpower, ease of training, ease of deployment and maintenance, risk of obsolescence, and so on.

It's important to keep track of all these through the iterations of the thinking and decision-making processes. If one has to re-visit the assumptions and the decisions later, it is vital to know and understand all the parameters and criteria that went into the process.

After going through the steps of defining one's identity and finding potential growth and opportunity areas, one can start setting the boundaries within which play of exploration and innovation can happen. The boundaries should be defined with clear reasons. Examples of boundaries include:

Doing only online business, addressing markets in certain geographies only, using largely open-source technologies only, sticking to three product lines only, targeting only a certain age-group, addressing only mid-sized companies, remaining a non-profit organization as mandated by regulations and hence clearly avoiding certain business models/practices, and so on.

Within the laid-down boundaries that are clearly defined with reasons, one should allow innovation to flourish unfettered in the organization. All assumptions other than the focus and the restrictions of known boundaries can be questioned, new insights welcomed, and collaborations forged.

This must be made clear across the rank-and-file of the organization. Not doing so is the single biggest weakness of any organization.

All the tools must be made available for freely accessing information within and without the organization, especially tools to filter, sort, and gather specific information. Of course, different grades of security measures should be in place to access increasing levels of sensitive information.

With the above measures and processes in place, high connectivity would then open up the entire organization to explore and innovate with a free spirit, but with relentless focus.

To drive the point home, a few simple examples are in order:

1. Consider an ace tutor in Math who specializes in pre-university level of subject expertise and holds local classes after school hours for students. He will find rich opportunities online if he remains focused in his area. This is because the larger crowd that is virtual and remote, and is in pre-university, will get strongly attracted to his deep Math capability, badly enough to want to connect with him. Using ICT, he can then schedule online virtual classrooms, as well as give and grade assignments through mail or shared folder. People can sign up on his website and contact him on his mobile phone. It would not do well for him at all to start teaching Physics also, or to even take up degree-level Math by rationalizing that "he is capable of doing it and because there is a large market". That would dilute his focus and reduce his value proposition. Not only will he run into much greater competition against other seasoned online/offline Physics tutors sooner or later, but he will drain himself of his time

and energy trying to master nuances in an unfamiliar field. That energy could have been better spent in sharpening his focus and value proposition, and marketing himself really well. Leveraging open innovation, he can partner with the likes of eBay, Monster, edX, and so on and also collaborate with other tutors and institutes. It's important to note that despite all the collaborations, and the final composite offering of a set of subjects to the student from such collaborations, he remains focused on Math. The collaborators, partners, and clientele also see value in him for the same reason.

2. *Lululemon Athletica*—Athletic apparel retailer. Founded in 1998, the 2017 figures as per Morning Star are revenues, US$2.3 billion; gross margin, 51%; operating margin, 18%; net margin, 12%. The revenues grew at 12% year-on-year.

   Compare this to Amazon, the largest Internet retailer, which is not at all as specialized as Lululemon. Its corresponding figures for 2017 are revenues, US$35 billion; gross margin, 35%; operating margin, 2.3%; net margin of a mere 1.3%. This is despite a 23% jump in revenues as compared to the previous year.

   Evidently, focus helps.

3. Examples of recent startups with deep and narrowly focused ICT products/solutions are:

   • Balance Eye (www.balanceeye.com) which is a VR solution using eye goggles that diagnoses balance and vertigo disorders.

   • TagBox (www.tagbox.in) which specializes in making cold chains more reliable. They monitor cold storage chain health in real time and predict and prevent failures, by using a combination of IoT, logistics, and mobile application solutions.

   • Imaginate (www.imaginate.in) provides a VR/AR/MR Collaboration Platform that is a B2B solution, which enables multiple people to communicate over AR/VR using minimal communication bandwidth of about 2 Mbps.

   • Docturnal (www.docturnal.com) detects pulmonary (lung) tuberculosis to the accuracy of 85% (currently) through a noninvasive method of analyzing the sound patterns of a patient's cough. It uses Machine Learning from past data to establish and refine its diagnosis model. In the future the startup plans to expand its scope to other lung diseases.

The Takeaway Box: Use High Connectivity and Unleash Forces from All Sides to Shape Your Focus Even Better!

- Use the process described earlier to identify the focus for your product.
- Define your focus based on the market needs, your uniqueness, and capability.
- Second, find different ways to engage and shape your focus in a myriad of ways.
- In the new unlimited world of possibilities, use frameworks and tools to maintain a sharp focus.
- Publish the clearly defined boundaries for the organization, thereby keeping the focus.
- Unleash creativity and innovation while leveraging the defined focus.

5.10   IF YOU ARE A STARTUP

Startups and small companies are hungry for profitability and growth. They badly need revenues and are often caught in the paradox of either chasing revenues in different directions and thereby getting defocused or sticking to one direction with a clear focus but having to tread a harder path and missing out on revenues in other areas. The answer lies in testing different areas, not through investments in full-blown product development but through discussions with different stakeholders, especially venture capitalists, potential customers, and partners, to find use-cases for the product. Zoom in on *one or two specific use-cases* only based on the following factors:

- The use-case aligns closely with the main/core/root functionality of the product.
- Ready customers available for the use-case.
- The single use-case must be a dire need for the customer, not a nice-to-have functionality.
- It's a low-hanging fruit for the organization relative to the product in question, with regard to the effort or investment needed for product development, marketing, and brand building.
- Large untapped (existing) market value.

- Very little or no competition (Blue Ocean strategy).
- Breaking into the market with one or two leading customers will build a pipeline of many more.

The market use-cases may be very different than what one had envisaged in the beginning, but it is important to allow the market and competition to play their role in defining the focus. The concept of "Pivoting" occurs when a startup realizes that its product is unlikely to succeed in the market with the initially dreamt of use-case and it has discovered a new use-case which has much better chances. The older use-case has to go, and the entire product development has to be aligned to the new use-case.

It is very dangerous for a startup to chase multiple use-cases, because it is extremely difficult to cut it in the market even while pursuing only one or two high-level use-cases (for the core product). Getting the first few customers is hard enough. It is strongly advised to build the entire brand, marketing, and sales around one or two use-cases.

As mentioned earlier, the Traction Gap Framework (Wildcat Ventures 2017) is a very useful framework that helps startups create the Minimum Viable Traction.

5.11   If You Are a Mid-sized or a Large Company

It is vital to retain focus over the years. And this is where most organizations are challenged. It is easy to get defocused in several ways, in the pursuit of growth, market share, or meeting quarterly sales targets. Guard against:

- Rampant product line extensions.
- Over-solutioning for a specific (usually large) customer, morphing the core product permanently.
- Collaborations that shift the center of gravity to outside one's core area.
- Fuzziness in operations and processes resulting from not defining one's focus in the first place, resulting in both wastage of resources and damaging the culture.
- Wrong hiring—getting people who do not really buy into the culture or the focus areas of the company. Apart from the other problems arising out of non-performance or HR issues, the key problem is that they may take short-sighted decisions that dilute the focus of the company.

5.12   Destinations for a Telco

A good example of discussion regarding business focus would be for a Telco to understand its destination in the ecosystem. (Creaner 2017) identifies five possible destinations for a Telco as described in the second row of the matrix below. The considerations are given in the second row.

#	Telco's possible destination	Considerations
1	Dumb pipe	Purely an underlying infra player. Scale, efficiency, and SLAs dominate. B2B where customers are large enterprises or other Telcos and media players. Huge infra investments into backbone networks. Large collaborations with a few players. Heavy on regulations, such as spectrum licensing.
2	Commodity player	Mostly B2B play, some business models with network and bandwidth packages, straddles both sides, with limited B2C play. Regional considerations. Heavy on infra, limited on marketing/sales. Big partnerships.
3	Smart digital pipe	B2B2C play that is closer to the end-customers, than the first two. Has visibility into end-customer needs. Typical clients are ISPs and OTT players, media companies.
4	IDSP (integrated digital services provider)	The most integrated B2B and B2C play that spans a wide range of digital products/services, from consumer access networks to enterprise broadband services. Many collaborations. Strong need for platforms. Strong competition, and need for innovation.
5	Services enabler/retailer	The topmost player in the pyramid. It's B2C or B2B2C and has no infrastructure, for example, an MVNO (mobile virtual network operator). Heavy on collaborations, marketing, and sales. Keen understanding of end-user business models, and striking numerous mutually beneficial partnerships with underlying CSPs on which it relies for its infra needs. Typically not too investment heavy as the others. Reliant only on platforms. Rampant competition and constant need for innovation. Margins can be high if the products are innovative and unique.

Based on the topics discussed in this chapter, a Telco can undertake a self-analysis, using the Business Model Canvas, as to its history, strategy, competitive positioning, partners, investments, ambitions, regulatory needs, and so on. A Telco may jump from position one to the next levels progressively over time depending on where it wants to be focused. It may choose to continue in one position forever.

5.13 CONCLUSIONS

Focus is key to success, thereby warranting an entire chapter in this book. There are numerous ways in which one can lose focus, and these happen naturally due to inertia and entropy, hence necessitating active action, frameworks, and processes to prevent this from happening. There are reference frameworks given in this chapter to help find, retain, and leverage focus. Simple processes built into the system and timely steps can help one achieve greater focus.

Before we move to discuss the strategic aspects of innovation, it is important to have a discussion on "keeping your focus right". It simplifies the strategy formulation process and clearly helps in developing strategy option. While developing ICT strategy, company requires to analyze its assets, capability, motivation for innovation, and so on, along with economic issues such as nature of demand, competition, and cost of production.

The Takeaway Box: Focus Simplifies Your Strategy and Execution

- Focus simplifies thinking, the decision-making, and the processes at every level.
- Prevention is better than cure—put measures in place to avoid dilution of focus.
- When in doubt, narrow the focus, and double down on the few critical use-cases for the valued customer base.

REFERENCES

Birudavolu, Sriram (2015): "Open Innovation in ICT: An Empirical Assesment of Global Telecommunications Services"; Unpublished Thesis submitted for PhD at Indian Institute of Foreign Trade, New Delhi.

Creaner, M. (2017, September). Telco Digital Transformation: The Conditions, Journeys, and Destinations. *Huawei*. Retrieved from http://www.huawei.com/en/about-huawei/publications/winwin-magazine/plus-intelligence/telco-digital-transformation.

Griffin, T. (2017). *12 Things about Product-Market Fit*. Retrieved from https://a16z.com/2017/02/18/12-things-about-product-market-fit/.

James, C., & Harney, B. (2012). *Strategy & Strategists*. Oxford: Oxford University Press.

Kim, C., & Mauborgne, R. (2005, 2015). *Blue Ocean Strategy*. Harvard Business Review Press, ISBN 1-59139-619-0 978-1-62527-449-6 (Expanded Edition).
Lance, V. (2016). Top 20 Reasons Why Startups Fail. *Forbes*. Retrieved from www.forbes.com/sites/groupthink/2016/03/02/top-20-reasons-why-startups-fail-infographic.
Maurya, Ash, 2018, Lean Canvas, Amazon, https://www.amazon.com/Ash-Maurya/e/B006MW5OSS.
NASA Tech Briefs. (2018). Retrieved from www.techbriefs.com.
Osterwalder, A., & Pigneur, Y. (2010). *Business Model Generation*. John Wiley and Sons, ISBN-10: 0470876417, ISBN-13: 978-0470876411
Ries, A. (1996). *Focus: The Future of Your Company Depends On It*. New York: HarperCollins.
Ries, E. (2011). *The Lean Startup*. Crown Publishing, ISBN: 0307887898.
Wildcat Ventures. (2017). *The Traction Gap Framework*. Retrieved from http://wildcat.vc/wp-content/uploads/2017/09/Traction-Gap-Framework-9.14.17.pdf.

## CHAPTER 6

The Innovation Strategy

Yuktih Shakteh Garisyasi (Strategy is better than Strength).
—Sanskrit Saying

6.1 INTRODUCTION

As per Peter Drucker:

> Strategic planning is the continuous process of making present entrepreneurial (risk-taking) decisions systematically and with the greatest knowledge of their futurity; organizing systematically the efforts needed to carry out these decisions; and measuring the results of these decisions against the expectations through organized, systematic feedback.

Hence strategy is not about tactics or operations, nor is it about forecasting or taking future decisions. It is doing analysis and making good decisions (in the present) about the future. It is commitment of resources to action. Owing to uncertainty about the future, it involves both risk-taking and innovation, in order to exploit opportunities.

6.1.1 Scenarios

Imagine yourself as the protagonist in the following mini-stories. Read through them and ask yourself what you would do in each situation:

© The Author(s) 2019
S. Birudavolu, B. Nag, *Business Innovation and ICT Strategies*,
https://Doi.org/10.1007/978-981-13-1675-3_6

1. You are the CIO tasked with the creation of an online job portal that brings together millions of job-seekers, thousands of employers, and employment agencies, some of whom are abroad in 50 different countries. You discover that the job-seekers need help with resume writing, sophisticated matching of their profiles to the different opportunities, and training and certification services to address skill gaps in both technical and soft skills, and even passport/visa/travel services. Furthermore, the job-seekers belong to different categories—freshers, experienced, senior, work-from-home mothers, retired, and so on—and they seek jobs that are full-time, part-time, consulting, and so on, all of which need different approaches. You also need to set up marketing and sales divisions, support centers accessible by phone and web, interface to other social networking sites, and so on. Your portal should also have a convenient mobile phone interface through a mobile app. Security, not merely confidentiality, is a prime concern as you store a lot of details about millions of people. There needs to be an effective feedback mechanism too through the portal and harvesting of social media sites for information.

   In short you need to build an entire ICT ecosystem and would like it to be innovatively different from the others in the market. From where should you start?

2. You work in the IT department of a pharma company that must essentially orchestrate information from different divisions inside and outside the organization—research labs, suppliers, a legal department that manages intellectual property, commercial, manufacturers (to whom outsourcing is done), partners, marketing, and sales. And there are also other departments like HR, recruitment, training, contracts, and so on. There need to be support and interaction channels with the end-consumers also.

   Of late it has become very challenging to integrate different systems and get information into them or out of them in a coherent and timely manner. Systems have grown over time and have increasingly become fragmented. There is a lot of expense merely to maintain the status quo functionality. It is getting almost unbearable in terms of effort and costs. Due to changes in business, there will be a load of new requirements in the coming year. You've heard that a lot of new technologies such as cloud analytics could be helpful.

   How will you move things differently this year?

3. You run the IT department for a metropolis. A smart city project has been mooted. There are a fleet of public buses and a train metro system which together provide mass-transit in the city. Due to traffic congestion and road construction/maintenance issues, there is a variance in the bus schedules at different times of the day. If the citizens are provided with a mobile app that shows them the updated estimated time of arrival of their bus at their bus stop, they can coordinate their schedules well, even synchronize them with the train timings at the metro station closest to the bus stop. They can also plan their pickup/drop/parking. This would save hundreds of millions of dollar-worth costs each year in terms of congestion costs, fuel wastage, man-hours lost, and so on. For doing this, the buses need to be fitted with GPS trackers. The central station needs to overlay the moving position of the buses on the bus route, which itself is layered on top of the geomap of the city.

There will also be constant dynamic feeds into the system from all over the city via messaging and phone calls regarding factors that may affect the movement of the bus, such as crowds or rallies, traffic congestion, road construction, drain works, rain flooding, and so on.

More train routes and additional buses as well as bus routes are being planned for the coming years. The city will also have more flyovers, roads, and expansion. All these will need to be considered. If the system is successful in this city, all the other cities in the country would like to port this system for their cities; hence this system needs to be built for re-use from the ground up. The system needs to be constantly improved based on feedback through various channels from the end-users.

Evidently, there is a need to bring together many ICT product and service innovations. How will you begin the whole program?

6.1.2   The Broad Questions

The scenarios listed in the previous sub-section reveal a real need to find a framework that one can use to implement an ICT strategy. Borrowing from the general co-creation framework by Ramaswamy and Gouillart (2010), into our discussion, the broad questions to ask are:

In the context of ICT, how are you:

1. Framing the experience mindset, and expanding the space of experiences?
2. Defining the context of interactions, and expanding the scope and scale of interactions?
3. Setting up the engagement platforms, and expanding linkages among engagement platforms?
4. Establishing the network relationships, and expanding stakeholder relationships in the ecosystem?

6.2   BASICS OF INNOVATION STRATEGY

Business innovation is at the core of the corporate strategy building. As mentioned earlier, companies are generally engaged in some degree of innovation both at technological front and in marketing of the products and services. The basic concepts of strategic management are broadly divided into three categories: strategic position, strategic choices, and strategy action. Innovation needs to be aligned with these for successful execution and commercial benefits. A summary of these three aspects are given in Table 6.1 below. At the initial stage, a company needs to understand the importance of ICT infrastructure and tools based on nature of market and competition, its resources, and competencies. At the second stage, it requires to do number of choices for optimizing its ICT tools by identifying products, understanding its customer value, and arranging

Table 6.1 Three pillars of strategic management

Strategic position	Business environment including state of competition
	Expectation and purposes
	Resources and competencies
Strategic choices	Corporate-level strategies (overall objective to fulfill stakeholders' expectation)
	Business-level strategies (choice of product, customer satisfaction, etc.)
	Methodologies to execute strategy
Strategic action	Managing change (cultural aspects, skill enhancement, etc.)
	Enabling (strategy to support particular effort, such as ICT enabling)
	Organizing (changing the structure of the organization to support successful performance)

Source: Modified from *Exploring Corporate Strategy: Text and Cases* by Gerry Johnson and Kevan Scholes; Sixth Edition, Prentice Hall of India, 2002

finance for innovation. Finally, it requires few changes at the organization level so that new products or services based on innovation are delivered to the customer successfully.

Suppose, the leaders or managers at the top have recognized the potential of an innovation and have agreed to move ahead, how would the firm ensure that it will be successful? It depends on how the firm is inclined to adopt the innovation. There are seven possible mechanisms to adopt an innovation.[1] It may be through internal development, internal ventures, or through licensing or acquisition or joint ventures/alliances. It also may be through venture capital-promoted nurturing and educational acquisition. It all depends on how familiar or unfamiliar the technology is. If the market is familiar and technology is known, the company may extend it through internal development as the risk is low. However, the more unfamiliar the market and the more unknown is the new technology, a company can think of partnership as its innovation strategy. In such cases, collaboration, joint venture, and so on can be the strategic choice rather than investing on its own and bearing the entire risk of innovation. This is popularly known as open innovation strategy (Fig. 6.1).

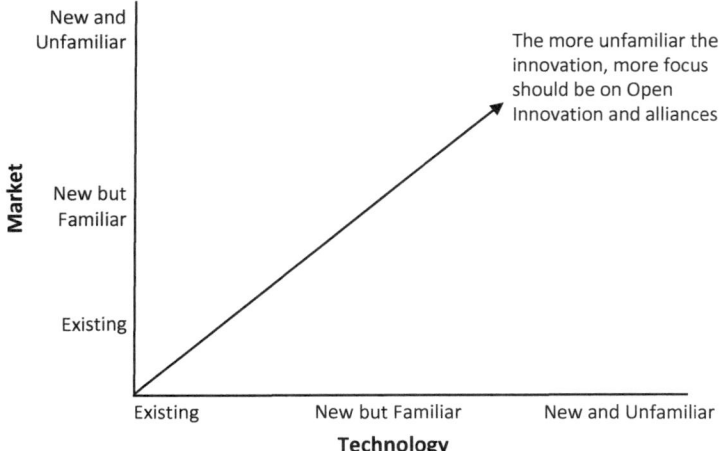

Fig. 6.1 Familiarity of technology and market: driver for open innovation (Source: Modified from Fig. 2.8, in *Innovation Management: Strategies, Implementation and Profits* by Afuah 2003)

[1] Afuah (2003): Models of Innovation (Chapter 2) of Innovation Management: Strategies, Implementation and Profits.

Historical studies on organizations which went through a structural change reveal that a firm maintains a long period of continuity with minimum change in strategy or the change is slow and incremental. A firm adopts new technology slowly, as long as competitive pressure is less and urge for innovation is subdued. There may also be a phase of flux where firms may adopt some innovation for strategic change, but there is not much clarity and direction. In most cases, firms mimic other firms and make attempt to play around the same technology. The innovations which bring a major transformation in strategy are not frequent. As described earlier, the strong force within and outside for a change pushes a company for a major shift. The following figure explains this in brief (Fig. 6.2).

Finally, there is a requirement of synergy between innovation and corporate strategy which depends on the match between two kinds of fit. Firstly, success depends on how far critical success factors and company's intangible assets such as skills, resources, and characteristics are well aligned. Critical success factors are those features which are valued by customer. A small grocery store while competing with big retailers focuses on personalized services, extended opening hours, informal credit, home delivery, and so on which are considered as major tools which grocer uses to overcome competition, and hence, they are the critical factors for success. Secondly, it also relates to how far new opportunities are well fit with

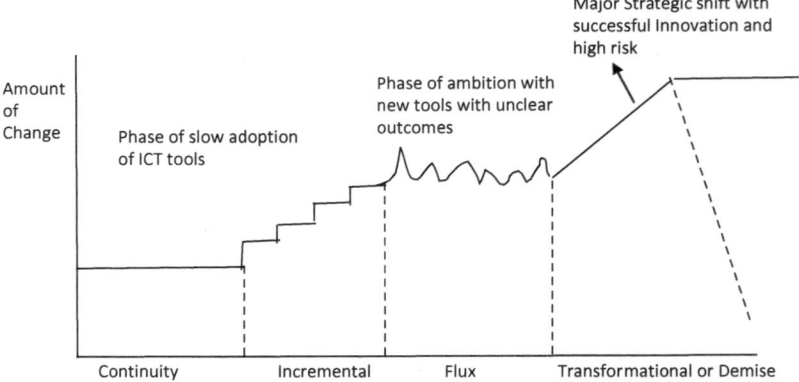

Fig. 6.2 Pattern of strategic shift (Source: Modified from *Exploring Corporate Strategy: Text and Cases* by Gerry Johnson and Kevan Scholes; Sixth Edition, Prentice Hall of India, 2002)

the capabilities of the firm. This is about benefit and opportunity. Is it an extension of the same business or completely a new business or doing the same business in a different way (by improving the product delivery and process reengineering)? A high fit in this case explains that value to be generated by innovation matches with the opportunities. When both fits are low, it implies that new opportunities and firm's competency do not match well and there is also a mismatch between firm's critical success factors and firm's resources. In such cases, innovation attempt will not add value to firm's business. Figure 6.3 below provides a snapshot of this. Sometimes, a good fit between new opportunities and firm's competency provides an attractiveness. However, this has to be vetted with respect to critical success factors related to firm's business. For example, a small grocer may find an excellent opportunity in selling specific products through an online platform, but it does match with his critical success factors. He will not be able to take any advantage of those critical factors (such as

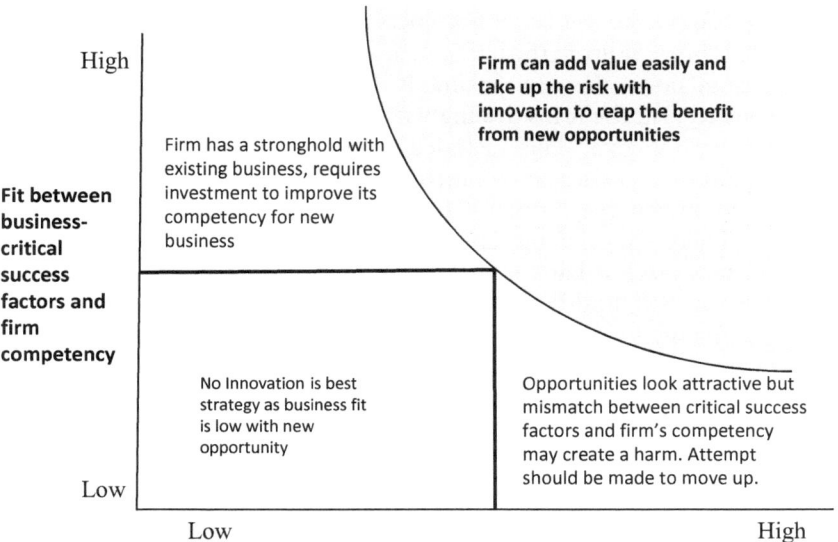

Fig. 6.3   Business fit and innovation strategy (Authors' own creation based on the Exhibit 6.11 from *Exploring Corporate Strategy: Text and Cases* by Gerry Johnson and Kevan Scholes; Sixth Edition, Prentice Hall of India, 2002)

personalized service, informal credit, etc.), and he has to compete with big and small players online. On the other hand, when firm is in strong footing with its business and it wants to explore new opportunities, the firm has to see whether it has sufficient resources, skill, and capability to scale up. If the firm is confident, then it can move ahead with investment.

6.3   Shaping an Innovation Strategy

Consider a few of the following events and trends that are causing an upheaval in the markets and economies. How should firms in the ecosystem respond to these massive changes in the market and regulations?

1. India moves towards a cashless economy in November 2016, with a view to:
    (a) Stem corruption, curb black money and illegal hoarding, stop counterfeiting and financing of crime
    (b) Strengthen banking, financial institutes, and the economy
    (c) Accelerate velocity of business
    This move spurred the immediate and large-scale adoption of online transactions, thereby making ICT central to trade, banking, and finance. All the firms dealing with cash transactions need to change their stratagem. This included banks, retailers, restaurant chains, transport services, e-commerce, and online vendors who collect cash on delivery, hospitals, and so on.
    There was a significant boost to all modes of digital payment systems: online bank transfers (NEFT/RTGS), mobile wallets (e.g. Paytm, SBI Buddy, Oxigen Wallet), credit cards, debit cards, and government initiatives like RuPay, UPI payment systems, and Jan Dhan Banking.
2. Explosion of sub $1 sensors, coupled with the adoption of IPv6 in networks (thereby facilitating nearly unlimited space of IP addresses), makes the IoT (Internet of Things) suddenly attractive. This can generate an unimaginable number of business models.
    Locanix (www.Locanix.com) is an IoT startup that uses GPS and ECU data sensing to provide innovative logistics solutions to do: real-time vehicle activity and location monitoring, smart fleet management and maintenance systems, driver behavior analysis, engine

health (RPM and temperature), fuel consumption, and impact/ accident detection.

3. Facebook shakes up the entire telecom/ICT ecosystem by creating open-source hardware and software on:
    (a) Telecom (Telecom Infra Project) (TIP)
    (b) Mobile networks, this will eventually become a part of TIP (OpenCellular)
    (c) Data centers (Open Compute Project) (Open Compute)
    Many major IT/ICT players have joined these forums, for example, Vodafone, Intel, Nokia, Cisco, Juniper, Accenture, Deloitte, MTN, Tata Communications, and so on.

4. The rise of the non-Telcos in the ICT ecosystem.
    Internet companies like Google and Facebook make aggressive strides into becoming Telcos themselves through launching projects like FB's Terragraph and ARIES, Google's Loon, SideWalk, and Fiber:
    (a) Or players like Apple, Oracle, and Microsoft heavily influence and shape the telecom/ICT landscape, with their devices, software, and cloud services.
    (b) Most companies turn digital to varying degrees, thereby shaping the data traffic, the demands on the network, and the kinds of applications and trends.

The changes and trends fall into one of these categories:

1. Large-scale *shift to digital services and to the relevant business models*:
    (a) Customer experience (CX) dominates and drives all the models.
    (b) Digital platforms to facilitate standardization, collaboration, innovation, and high scalability.

2. Shift towards *software*—the network, the servers, service stacks that provide services, even the services themselves, the economy, and so on. Even "hard goods" are a product of software and are heavily dependent on software for their specifications, design, production,

logistics, usage, maintenance, and so on. The important points to note are:

(a) Developer experience (DX) becomes as important as CX.

(b) This is to facilitate a large community of different types of software developers to produce rich, complex, and scalable software faster, cheaper, and without errors.

3. An explosion of new and *low-cost technologies*, both software and hardware:

(a) Examples are open-source software, cloud computing, low-cost sensors, servers, network devices, and so on.

4. All the above result in *data becoming extremely valuable*; it is a *data-centric world*; thus analytics, data science, and modeling and mining assume importance as never before:

(a) Increase in scale, variety, and speed of data, hence the need to handle the so-called 10Vs of big data (Firican 2017):
volume, velocity, variety, variability, veracity, visualization, validity, vulnerability, volatility, and value

(b) Small data becomes all too important. While big data represents trends that have already taken shape, small data represents nascent trends and tiny clues that represent future possibilities, such as untapped markets, outliers, or even negative scenarios such as hint of fraud, security breach, or good practices vanishing (which need to be protected and increased urgently).

(c) Data security has become a prime concern.

Examples of analytics platforms are:

- IBM Watson (www.ibm.com/watson/)
- GE Predix (https://www.ge.com/digital/predix-platform-foundation-digital-industrial-applications)
- Datamorphix (www.datamorphix.com) (Fig. 6.4 and Table 6.2)

5. Increase in *collaborations of every kind and open innovation*—alliances, partnerships, forums, and open sourcing/crowdsourcing (which are also a form of collaboration on a large scale):

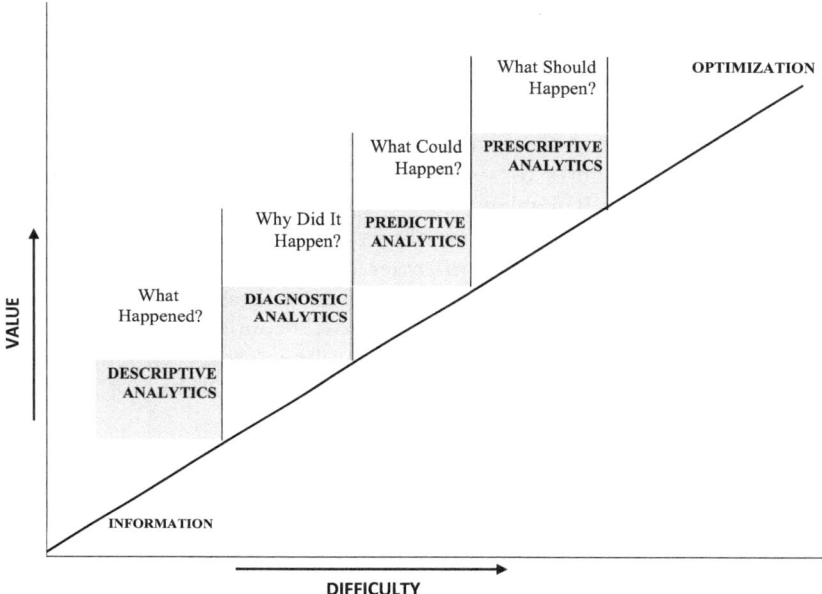

Fig. 6.4 Analytic value escalator (Source: Intelligencia Limited, Intelligencia (2018))

Table 6.2 Details of the kinds of analytics

#	Type of analytics	Description (examples)
1	Descriptive	Use Google Analytics to track website traffic (page views, visitors, etc.)
2	Diagnostic	Mining the data to determine why there was a spike in web traffic last week
3	Predictive	Using algorithms and data to predict which hotel customers are very likely to use a specific offer
4	Prescriptive	Using optimizing algorithms to optimize utilization, hotel room rates, costs, workforce

Source: Intelligencia Limited (Intelligencia 2018)

(a) Solutions that integrate best-of-breed ideas and components
(b) Specialization, experimentation, and rapid improvements favored
(c) Speed and time to market are a priority

(d) Need to be built into the vision, strategy, architecture, and implementation; should not be considered as an afterthought or attempted to be retrofitted

Topcoder (www.topcoder.com) is an example of a platform that brings together over a million software developers, designers, and technology experts to solve complex ICT problems. It is about crowdsourcing ICT talent. Other types of forums and collaboration are discussed in later sections of this chapter.

6. Rise in importance of *architecture (both top-down and bottom-up) and frameworks*, the emphasis being on antifragile, scalable, open, standards-based architectures with loosely coupled components.

Examples: TOGAF for Enterprise Architectures, TeleManagement Forum's Frameworx for ICT businesses, IEEE 3030 by C3DP for 3D printing.

7. *Stronger regulations and industry practices*, in every sphere, that spur digitalization, security, competition, open systems, innovation, growth in business/trade, better governance, sustainability, and elimination of wastage.

Examples: data security regulations specific to each country such as the European Union's GDPR, open-source O/S Linux for PC, Android for mobile, GSM mobile communication standards, HIPAA for medical data, OWASP for web security, and so on.

All of these seven categories, or dimensions listed above, are obviously interrelated and influence one another. Evidently, any innovation strategy for ICT must necessarily account for each of these dimensions. The different dimensions of ICT innovation strategy are summarized in Table 6.3.

From this table, it is quite clear that many major changes are underway, and firms can view these changes as opportunities to take their business to the next level. The main principle is to move from a purely product/services mindset to an experience- and innovation-oriented thinking. The coming sections of this chapter will discuss the ways to shape the firm's ICT strategy based on innovation, especially open innovation.

Table 6.3   The dimensions of ICT innovation strategy

#	Dimension	Salient features
1	Shift to digital services	• Customer experience (CX) • Digital platforms and standardization in protocols and interfaces
2	Software powered	Entire software ecosystems; reduce hardware in favor of software 1. Developer experience (DX)
3	Low-cost technologies	• Open source-based software and hardware • Diversity of devices and software made possible due to standardization of protocols and interfaces • Cloud based to lower cost, hide complexity, and increase utilization of resources
4	Data-centric world	• Big data (10Vs)—volume, velocity, variety, variability, veracity, visualization, volatility, validity, vulnerability, and value • Small data • Data security
5	Collaboration and open innovation	• Partnerships, forums, alliances, crowdsourcing • Driving principles: specialization, best-of-breed ideas, and speed
6	Architectures and frameworks	• Integration is key, open standards based • Antifragile, scalable • Layered and segmented; separation of business process from implementation • Re-use, modularization, and loose coupling of components • Principle of minimalism • Processes and metrics are an integral part of the architecture, especially agile processes
7	Regulations and industry practices	• Governance • Competition • Open standards • Digitalization • Security • Sustainability • Elimination of waste; increase utilization of resources, for example, IT, network, and software development resources • Improve business, opportunities, living standards, and the economy

Source: Birudavolu (2015)

6.4   Open Innovation and the ICT Landscape

As per the study done by the authors (Birudavolu et al. 2016), globally, the different upheavals and problems that not only Telcos but all firms dependent on ICT need to contend with today are:

> Hyper competition, constantly rising expectations of customers and all stakeholders, downward pressure on prices/revenues, guaranteeing ROI for stakeholders, meeting regulatory demands, constant changes in technology and markets, increase in media and communication devices such as smartphones & smart devices. (TeleManagement Forum)

Deep insights regarding customer experience, high degree of automation, and constant innovation are the way to success. Organizations need to invest in innovation continually to find new solutions constantly, or they will fail (Christensen 1997; Christensen and Raynor 2003). Hence firms focus and produce value in their core areas while collaborating in all others that are not its chosen domain. The firm forms partnerships outside its boundaries to deliver high value to the market, as shown in Fig. 6.5 (Chesbrough 2003). Open innovation is defined as innovation driven across the boundaries of an organization (Chesbrough 2003), in both directions.

The firm exposes its unutilized assets to other organizations which lie outside its boundaries, so that it can find a way to monetize the assets while others derive value out of them.

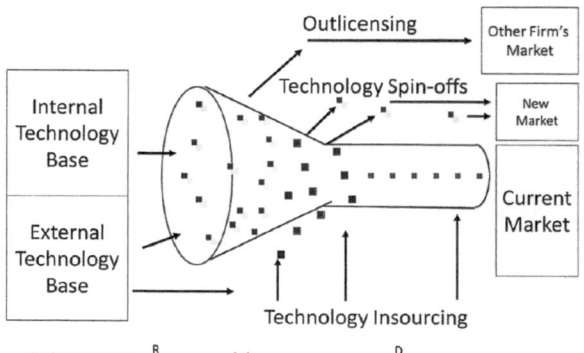

Fig. 6.5 Open innovation depiction (Source: Chesbrough 2003)

Thus, open innovation is about consciously scanning a wide range of internal and external sources for innovation opportunities and selecting and integrating the chosen sources with the firm's own capabilities and resources, to exploit market opportunities through multiple channels (West and Gallagher 2006; Bigliardi et al. 2012).

At the traditional end of the spectrum, closed innovation is done purely inside the organization using its own R&D resources, and the intellectual property produced is then sifted, sorted, and used by the marketing and production divisions of the company (Chesbrough 2003) to roll out products and services. In this case, the organization entirely bears the risks and rewards. Closed innovation is limited by the organization's resources, that is, the R&D budgets, its talent pool, domain expertise, its response time to the changing conditions in the market, the risk appetite, the marketing power, hierarchy of the organization, technological capabilities, and so on. Open innovation is a way to overcome the limitations while augmenting the native strengths of the organization's closed innovation. As an example, active collaborations with external content/service providers, related to the core/growth areas of the Telcos, represent open innovation.

Open innovation helps reduce time to market, lower integration costs, converge services, and provide competitive pricing on existing services (Bigliardi et al. 2012). The ICT industry needs to rely on ideas and technology from external partners (Nesse 2009). The search for strategic innovation is the key factor that is a hallmark of companies in the ICT industry (Bigliardi et al. 2012).

Examples:

- Yes Bank (www.yesbank.in) provides collaborations in the form of a "Bank from ERP" solution using platform with API (application programming interface) to allow businesses to access a third-party system for performing banking operations straight from their own ERP (enterprise resource planning) systems. This enables secure access to their digital assets and services. All payables, receivables, cards, card-less withdrawals, utility payments, and channel (dealer or vendor) finance can be managed using Bank from ERP.

- Arkadin Cloud Communications (www.arkadin.co.in) is a subsidiary of NTT Communications that provides cutting-edge ICT infrastructure. For enterprise consumers it provides and facilitates audio/web/video conferencing, virtual events, and unified communications. With this it collaborates with over 50,000 businesses in 30+ countries to provide them premium service. Many tiers of Arkadin services incentivize collaboration within and across businesses, as a result of which Arkadin itself benefits from the collaborations.

6.5   Keys to Open Innovation-Based Strategy

6.5.1   *The Global Innovation Index and What It Means for You*

The intensity of open innovation in an organization is a function of the open innovation strategies being put into practice by the top management, and the open innovation social capital built in the organization over the years, both of which give the organization open innovation capability in the market. An ICT firm's strategy is shaped by the environment it operates in. Environment is important for innovation. Any innovation happens in the context of a socio-political, economic, development, regulatory, infrastructural, and technical background. Open innovation is no exception. Hence, for emulating any global practice, it's important to factor in the country's effect on the strategy. The country's basic innovation capability is determined by several factors, such as the infrastructure, the level of education of the population, the governance, and so on. A strategy that has succeeded in Germany may not work in Uganda, and vice versa. To do this, especially for launching new services in a country, one may consider using the GII (Global Innovation Index) from the World Economic Forum, which considers a set of 40 variables and computes the innovation index per country. If the GII of ones' country is lower as compared to another nation's, and one is attempting to replicate a digital service of that nation in one's country, then the entire service model needs to be thoroughly re-examined. The ICT firm needs to determine and tune the technology, mode of deployment, partners, channels, pricing model, marketing, operations, service, support, and so on for that country.

You need ongoing primary data too, relative to your product/service, to validate the specifics as against the Global Index parameters. This is because the mileage varies from area to area even within a country. Well developed areas in a backward country may have higher connectivity and educated people, and it may make all the difference to your business. Whereas there could be less developed, far flung regions within an economically/technically advanced country.

Examples:

1. Staqu (www.staqu.com) provides AI/ML powered tools to help provide contextual recommendation to the users of your products/ services. It can understand users' behavior towards e-commerce, depending on the category, brand, vendor affinities, and other nice parameters, all of which are used to make recommendations to the user. It also has software sitting on the users' mobile app to understand users' content consumption in real time and help the organization address the user needs better.
2. InsideView (www.insideview.com) is a leading market research company that has tech profiles of over half a million businesses worldwide. It provides a searchable database of current companies and contact information, and real-time insights on the market happenings pertaining to companies, such as news, leadership changes, new offerings, acquisitions, deals, regulatory filings, litigation, and so on. It provides over 40+ fields of company and contact information and other details. Using tools, you can integrate their platform with your application to do further analysis and market research for your needs, for example, competitive analysis, targeting for marketing/ sales, collaboration and partnerships, understanding the kind of markets they serve in which region, and so on.
3. Tracxn (https://tracxn.com) is a database and a platform pertaining to startups' data. It tracks over 30+ million companies globally, across dozens of domains and geographies. It provides sectoral reports in different sectors and technologies. It is useful for tracking technology and business trends, the way the landscape is shaping up with startups in different geographies, searching for collaborations, investments or acquisitions, and so on.

The Takeaway Box: Conduct a Thorough Study and Build a Knowledge Base

- *Make requirements engineering* and *market research* a deep practice.
  - Explore thoroughly the entire end-to-end customer experience, that is, region-specific, country-specific, and person-specific.
- Document and imbibe the customer journeys, the use-cases, the odd cases, the problem scenarios, the touch points, and all the interactions.
- Document and understand thoroughly the behavior, preferences, the habits, tastes of all the key stakeholders at the different touch points in the ecosystem. The stakeholders could constitute diverse groups ranging from customers, end-users, to developers, maintainers, and operators of the system.
- Learn and document the behavior, problems, limitations, constraints of *the existing systems and technologies,* and the impact of these on the users.
- Understand thoroughly the *reasons* behind all the above—these could be due to regulations, or high cost of living, or cost of high living, regional factors, and so on.
- Study the *competition* thoroughly. These need not be directly competing products or services but could be something else altogether. Even unrelated things could be taking away the time, money resources from the needed deployment, because of habits or distractions. For example, the competition to studying a topic online needn't be merely another educational software but could be a cinema being shown on television that is taking away the student's time.
- *Re-use* heavily from everywhere—other countries, regions, competing products, and so on.
- Ideate regularly—study parallel systems and processes in ICT and other industries/fields and discover ideas. Document the ideas thoroughly in the searchable knowledge base.

6.5.2   *The Revenue-Sharing Arrangements*

However, the most important component of an ICT firm's open innovation strategy is indicated by the revenue-sharing arrangements with its ICT partners. The firm's ICT partners are its collaborators with whom it plans to launch services jointly. Comprehensively, this indicates the importance the firm attaches to the partnership, in terms of monetary value it is willing to share, out of the revenues generated by the services launched by the partnership. Not surprisingly, the revenue-sharing arrangements of Telcos/ICT vendors within a country are similar for similar kind of services, that is, they fall within a narrow band. This is because revenue-sharing arrangements are the easiest part to copy from each other over time, and the ICT/data services vendors also have bargaining power within a country—all of which has an equalizing effect on the revenue-sharing arrangements of the different firms in the country. This is due to precedence established through the cycles, and the benchmarks in the industry. There could be differentiating factors, such as additional security provided by one service provider, but the perceived values of the basic services converge. It is therefore essential to calibrate and measure any service being provided, in the context of the ecosystem, project appropriate monetary values based on the markets, and arrive at revenue-sharing arrangements based on mutual dependence with collaborators/vendors. This will be driven by the estimated value, effort, risk, reward, and the market benchmarks for revenue-sharing.

The important contribution of the research by (Birudavolu 2015) is the empirical evidence from the ICT industry that supports the theory of open innovation and thus helps promote its use in ICT firms. Competition within the ICT industry in a country is also shown to boost the power of open innovation as measured by the latter's effect on the firm's performance.

Revenue-sharing arrangements among collaborators are a vital part of open innovation strategy, and hence must be taken very seriously. The arrangements are indicative of the agreed value of the worth of each collaborator in the partnership. This share is not a passive investment game, or even an "outsourced" R&D arrangement, but is an active collaboration in which each party, the firm, and the MDS provider work together at each step throughout the process from planning to marketing, deployment, and operations.

At the very least, each collaborator should:

• Play an active role in some of the crucial steps of the process, and
• Participate in sharing risk, and
• Be included in commensurate reward sharing—this is not just a monetary relationship, but the collaborators share closer relations in terms of ideas, technology, human capital, and co-opt for bigger challenges

The motivation behind the collaborations should be:

- Bringing together the best-of-breed ideas, talents, and technologies
- Each collaborator brings to the alliance a crucial expertise that the other collaborators don't have
- Ideally, the collaborators don't have any other motivation for the alliance

A popular example is that of Apple and its iTunes ecosystem. An independent firm or even an individual may develop an application, adhering to the iTunes guidelines, and launch it worldwide on the iTunes platform. Apple keeps only about 15–20% of the revenue that the application attracts worldwide and gives away all the rest to the developer of the application. This revenue-sharing arrangement alone has been instrumental in the phenomenal success of the iTunes ecosystem, and consequently the tremendous success of Apple's products. Similar examples abound in Japan, Korea, and Europe where the collaborative ecosystems have been an explosive success.

Another example:

KDDI Japan launched a subscription program, AU Smart Pass, for developers in 2012, instead of launching an app store and competing with Google Play and Apple Store. The Smart Pass comes pre-installed on smartphones, so the developers can bring apps into the Pass, and interested subscribers just pay and unlock the app's features. KDDI gives away 80–90% of the revenue to the developers. As of December 31, 2017, the AU Smart Pass had logged 15.4 million users, of which there more than 3 million premium members, with monthly fee ranging from JPY 400 (normal price) per month to JPY 148 (discounted) per month (KDDI Financial Reports 2018).

Countries like India are yet to witness this level of success because the revenue-sharing arrangements are at a much lower level. This provides meager resources or incentives to the collaborators of the firm to provide richer services and constantly experiment and improve upon them. Consequently, they struggle for profitability and growth, thereby rendering the market and ecosystem confined to a handful of business models and a few players only. There are several examples of the ICT firms themselves taking on the business of providing a variety of digital services, through mergers and acquisitions or through diversification and organic growth, that is, in-house development. But this is an ineffective strategy in the long run because it does not tap into the diversity of the markets, society, and economy, and ends up stunting the entire ecosystem. Newer, smaller, talented, and agile players would be prevented from entering the market because of the steep walls of competition and entry barriers that the incumbent ICT would set up to prevent disruption to their existing digital services businesses. The incumbents would suffer stagnation and become inflexible and expensive. The consumers and the entire economy would thus suffer. This situation is changing now, with the Indian Telco players signing up more collaboration partners.

Here are a few examples of the collaborations from just one Indian Telco, Bharti Airtel:

- Airtel joins global consortium to offer in-flight connectivity (NDTV 2018).
- Airtel launches open network, in which Airtel has opened up its entire mobile network information across India through an online portal, which will display Airtel's mobile network coverage/signal strength and network site deployment status. Customers will get a clear picture of their own location's status and can raise complaints and get status as well (Airtel Open Network 2016).
- Airtel announced a digital innovation program, Project Next, to transform customer experience across all its touch points, which will allow it to collaborate very closely with its customers to improve its experience, for example, the Family Promise will allow customers to create customized solutions for their families in once click of an app (Telecom Drive 2017).
- Airtel ties up with Samsung to provide smartphones to its customers (Airtel Samsung Collaboration 2018).

A telecommunications industry analysis by the World Economic Forum in 2017 in collaboration with Accenture (WEF 2017) has described the possible revenue-sharing models for a Telco:

1. *Direct Partnership*: Telco does B2B collaborations in vertical market use-cases to launch services. Typically, it's a revenue-sharing model. For example, Airtel or Vodafone ties up with VAS (value-added services) players.
2. *Investments*: Telco makes investments in B2B collaborations (for profit share or stake) in several vertical market areas, thereby gaining

an insight into those business models. It can play a bigger role akin to a venture capitalist in finding synergies and having subsidiary companies collaborate or merge. For example, Bharti Airtel owns stake in Telemedia ventures.

3. *Community Approach*: Telco engages in B2B or B2C community collaborations for building business models in vertical markets, licensing, or revenue share. For example, Vodafone participates in many communities ranging from healthcare to women welfare.

4. *Platform Approach*: Telco engages in building B2C or B2B platforms that enable large-scale collaborations that will grow due to network effects, licensing, subscription, revenue share depending on the situation. For example, BT (British Telecom) has built a platform called BT PCMS (Personalised Compute Management System) that offers pre-built "Business-As-A-Service". It allows companies to launch their products and services on this platform.

The Takeaway Box: Overinvest in Partnerships

- The reward-to-risk ratio must be commensurate in partnerships and must be better than the market.
- Perfect the collaboration model and processes.
- Find ways to make the partnership grow and become successful.
- Partnership models should be inbuilt into vision and strategy and should not be considered as an afterthought.
- Open innovation collaborations are different from ordinary partnerships. They are not a typical outsourced-for-a-fee model—the OI collaborators jointly launch services and share both the risks and rewards.
- Understand the limitations of each partner and continue to scan the landscape for new, promising partners.
- With addition of each partner, the firm must increase the experience space being provided to the customers and end-users, and improve the richness of understanding.
- Collect metrics and benchmarks from all the customer/market interactions made possible with the partnership. This will serve well to improve market intelligence, and expand markets in other areas, as well as rely on other partners in future, should the current partnership fail to work out.

6.5.3 The Platform Approach

To vastly scale up a number and variety of collaborations in a frictionless manner, the platform approach is highly recommended. In this model, the firm constructs (or borrows and re-purposes) a digital platform that is suited for the needs of its domain. This platform is the lynchpin of its entire ecosystem. Collaborators can then build services that neatly plug into this platform and provide value to the end-consumers and/or the other players in the ecosystem.

Through the platform it is then easy to accomplish the following:

- Collaborate.
- Enforce regulations.
- Cater to architectural standards and frameworks, improve integration, and provide flexibility for future needs and growth.
- Mandate security measures and policies.
- Provide scalability and reliability—for example, adding servers and resources to the platform can take care of increase in volumes of users who consume a specific service, or adding resilience and disaster recovery to the platform can increase the reliability of a service.
- Improve utilization of resources; eliminate fragmentation, wastage, and complexity.
- Simplify procurement and deployment of resources (hardware, software, network, manpower/expertise).

Each collaborator then need not worry about these risks and overheads. They're all provided to each collaborator in a much more standardized and subsidized manner. The collaborator is focused on his core service to make it highly successful and profitable, through repeated iterations of tweaks and launches.

Currently firms are using open innovation as a means of de-risking their strategy even while being opportunistic about launching new services in the market. This has come about because the MDS (mobile data services) partner, who has the domain expertise and implementation capability, typically desires to try out new services in the market but lacks the means to launch those services. The firm, on the other hand, has the infrastructure and wherewithal to market and launch the services. The firm does not stand to lose much by extending its infrastructure marginally and collaborating with the MDS partner to jointly launch new services. The Telco/ICT firm also has all the tools and instrumentation in place at each level to gather detailed information about the launched service, the customer behavior, and the necessary analytics from every layer of the channel and stack. If the launched service fails in the market, the firm does not stand to lose much, but can in fact provide detailed feedback, to the MDS partner. Using this feedback, the partner can improve, tweak, and refine services and re-launch the services, or combine with other services or even shelve it and move on to another service with a different business model. If the launched service is successful in the market, then both the MDS partner and the Telco/ICT stand to benefit immediately. They can use the detailed analytics pertaining to the service provided by the firm to be able to refine and scale up the launched service. They can monitor every aspect of the service and progressively keep launching new improved versions of it in the market. Platforms are studied again in another chapter also.

Example 1

LUCY is a sustainable environmental monitoring IoT platform (LUCY 2018) from ThinkPhi, which is a clean technology startup. It consists of a sensor device that is deployed onsite in the field to monitor environment variables (gas/pollutant parameters) such as particulate matter (PM 2.5 and PM 10), carbon dioxide, temperature, humidity, and noise.

The key features of LUCY are:

- Scalability due to its ability to integrate with third-party products
- Self-sustainable data monitoring device
- Cloud-based management system and data connectivity
- Rugged build for all environmental conditions
- AI-powered data management and analytics

LUCY has a partner ecosystem, in which offices, residences, campuses, local governments, blocks of the city, and corporations can collaborate to deploy LUCY to monitor their environment for safety, health, and compliance. Through its partner network, LUCY can help its partners become successful with their clients. The partnership extends from installation, warranty, support (training, spare supply, etc.), to customer service. It is easy to produce reports, compare data with other regions, check data for regulatory compliance, take prompt action, and so on.

Example 2

Razer is a Chinese gaming company that makes half a billion US dollars in revenue and has a market valuation of about US$4.7 billion (Yahoo Finance Razer 2018). As of 2018, it serves 35 million users with just 700 employees! This is due to the platforms and processes it has put in place.

The Takeaway Box: Build Your Platform Immediately, and Base Your Ecosystem on It

- It is neither easy nor advisable to construct a platform from scratch.
  - Avoid constructing platforms and solutions yourself—leave it to the experts. There are too many parameters to cater to, including technology upgrades and keeping the solution up-to-date with evolving requirements.
- Collaborate and borrow technology platforms, products, tools and re-use heavily to build your ecosystem-based on the platform.
- Thorough requirements engineering and market research must precede the undertaking of building a platform and ecosystem.
- Do not attempt a large-scale transformation program to move everything to the new system. It is practically doomed to fail.
- Have legacy systems also plug into the new platform and then upgrade old functionality piece by piece to the new system.
- Measure the platform in terms of the utility and benefits it brings to the ecosystem. The benefits have already been listed earlier in this sub-section.

6.6   DIGITAL SERVICES FORUMS

6.6.1   Forming Collaborative Forums

The term forum is used in this book to mean multi-party alliances or collaborations. Innovation in digital services is core to the ICT companies' profitability and growth. Open innovation-based forums that cater to digital services and the ICT industry need to be set up. The ICT industry has recognized the potential and is investing in these alliances. There has been a steep rise in the number of ICT-related open innovation forums, hubs, and consortia, across the boundaries of the Telcos. To launch more innovative and profitable services, firms need to collaborate with a variety of partners, such as:

- Providers of mobile digital services, mobile data services, value-added services
- Software development and IT organizations
- ICT service providers
- R&D organizations
- Academic institutions
- Equipment vendors, OEMs
- Government bodies
- Competing ICT firms, in areas of mutual interest

Digital services encompass manifold domains, most of which are outside the remit of traditional Telcos, such as healthcare, media, agriculture, banking, government, regulatory, security, legal, advertising, publishing, retail sector, manufacturing, logistics, software product development, and so on. Here are about 40 extant open innovation ICT forums, sampled globally in 2015. These are shown classified along different dimensions, to underscore the diversity.

Open innovation in the free marketplace is growing strongly, and every organization should include it in its strategy. A firm must study its markets and industry and seek to set up or participate in appropriate forums that leverage the use of ICT in its business (Tables 6.4, 6.5, 6.6, 6.7, 6.8, 6.9, and 6.10).

Table 6.4 The dimensions of ICT strategy: classified based on the size of collaborators

S. No.	Size of collaborators	Examples
1	Large	EIT ICT, Holst Center, China-Finland Alliance, etc.
2	Medium	Adastral Park, Synergia, Fing, etc.
3	Small	iHub, KINU, FOSS4G

Source: Birudavolu (2015)

Table 6.5 Classified based on investors

S. No.	Origin	Examples
1	Government backed	Miriade, Lindholmen Park, Fing, iCluster, etc.
2	Corporation backed	Cisco EIR, AT&T Foundry, KDDI Open Innovation Fund, etc.
3	Funded from diverse sources	iHub, KINU, FOSS4G, etc.

Source: Birudavolu (2015)

Table 6.6 Classified on a geographical basis

S. No.	Geographic region	Open innovation ICT forums
1	North America	Cisco EIR (USA), iCluster (Mexico), TR Labs (Canada), etc.
2	South America	STI (Chile), Telefonica Innovation Hub (Brazil), etc.
3	Africa	ActivSpaces (Cameroon), MEST (Ghana), iHub (Kenya), HiveColab (Uganda), etc.
4	Europe	Open Living Labs, Lindholmen Park (Sweden), Adastral Park/Martlesham (UK), Miriade (France), etc.
5	Asia	Init (India), THTI (China), FOSS4G (Thailand), telecom ideas (India), etc.

Source: Birudavolu (2015)

Table 6.7 Classified based on commercial/non-commercial interests

S. No.	Commercial/ non-commercial	Examples
1	Non-commercial interests	GSMA OneAPI, FOSS4G, STI, etc.
2	Commercial interests	Plug and play, AT&T Foundry, KDDI Open Innovation Fund
3	Mixed	Fing, OpenAlps, Holst Center

Source: Birudavolu (2015)

Table 6.8 Classified based on Telco and non-Telco organizations

S. No.	Telco/ non-Telco	Examples
1	Non-Telco	FOSS4G, STI, Lindholmen, Fujitsu Labs, KLab
2	Telco	AT&T Foundry, KDDI Open Innovation Fund, Verizon Center, Telefonica Hub

Source: Birudavolu (2015)

Table 6.9   Parameters to rate the open innovation ICT forums

S. No.	Parameter	Metric description
1	Benefits—inbound-outbound	Inbound to outbound (1–5 scale, 1 being lowest). This describes whether the benefits are more inbound (i.e. acquiring ideas), or more outbound (sharing). (Elmquist 2009; Dahlander and Gann 2010)
2	Benefits—pecuniary	Relates to whether business or monetary gains are a part of the goals. Rated on a scale of 1–5. (Dahlander and Gann 2010)
3	Benefits—directness of use (1–3)	The three levels are symbolic (lowest), conceptual (medium), and instrumental (highest)
4	Locus of collaboration (1–5)	This ranges from internal (lowest) to external (highest), indicating whether the collaboration is within internal divisions/subsidiaries of an organization or whether it extends to a number of parties external to the organization
5	No. of countries	The number of countries that the collaboration spans
6	No. of orgs involved	The key strategic players in the collaboration
7	No. of labs	An indication of the research involved
8	Size of collaborators	Graded from very low to very high (1–5), where a startup company would be rated as very low and a large corporation or a government would be rated very high (Elmquist 2009)
9	Target (incremental—radical)	About target innovation—ranging from incremental innovation to radical innovation on a scale of 1–5. Pelz (1978)
10	Role of company (1–4)	Ranging from just being an investor (L), to a facilitator, idea generator, developer of platform (H) (on a scale of 1–4)
11	Organization structure (1–4)	Rigid teams (low), task forces, federated, mass collaboration (high) (on a scale of 1–4)
12	Type of collaboration (weak-strong)	From weak ties (low) to strong collaborations (high) (on a scale of 1–5)
13	Method of innovation (1–3)	Lead user method, ideation contest, mass collaboration (on a scale of 1–3) (Erkens et al. 2013)
14	Measurement type (1–4)	Innovation measurement scale: input (low), process, output, outcome (high) (on a scale of 1–4). (Erkens et al. 2013)
15	Years since established	Longevity of the forum

Source: Birudavolu (2015)

Table 6.10 Innovation matrix

		External (business)	Mixed	Internal (people)
Commitment	*Structural*	Startup fund	Emerging business areas	Center of excellence
		External incubator	Internal accelerator	Community of practice (ambassadors)
	Non-structural	Innovation challenge	Design Sprint	Innovation workshop
	Capabilities			

Source: Pirenne (2017)

6.6.2 Studying Characteristics of ICT Forums

This section helps a firm measure and monitor its level of participation in ICT forums, for shaping a relevant strategy, implementing, and governing it. These parameters ultimately boil down to the following factors:

1. Open innovation intensity in the collaboration.
2. Global footprint of the collaboration.
3. How established is the collaboration?

These will be used in the next section to elaborate on how a strategy can be formed.

> **The Takeaway Box: There Is More to Collaborations Than Meets the Eye**
>
> • Most people and firms do not really know how to form or use collaborations.
> • Only now are organizations realizing the power of collaborations, and hence the increase in the number of forums.
> • You must measure your collaborations.

6.6.3 Learning from the Industry Forums

Open innovation in ICT is being widely adopted in the ICT industry. As the study (Birudavolu 2015) shows, ICT and non-ICT organizations from different backgrounds are collaborating in forums towards creating and launching innovative digital services. In the open innovation alliances being set up, corporations (Telcos and non-Telcos), governments, research institutes, and non-profit organizations are funding startups. This reveals that there is immense potential in the space of open innovation in digital services, as the investors expect good profitability and growth from the startups that they're funding.

The study also shows a distinct inclination towards radical innovation rather than plain incremental innovation. It depicts that the more radical innovations come from companies that are a bit more established than the new and small ventures, as indicated by the mean age of the venture in years. This is because the firms have worked their way out of teething problems, in terms of getting funding, harmonizing the business model with the markets, and stabilizing the processes. The firms pursuing an open innovation strategy also form global alliances and collaborate with an increasing number of organizations. Their locus of innovation also moves more towards outside the firm. However, after firms reach a certain maturity, in terms of age, size, partnerships, global footprint, processes, and so on, the internal and external innovation are more balanced. Hence the firm finds that both incremental and radical innovation yield value. Hence a healthy mix of incremental and radical innovation becomes the mainstay of the forum at that stage. The firm achieves a good balance between open and closed innovation over a period. Ideas from open innovation spur better closed innovation, and shortcomings in closed innovation or misalignment of closed innovation with the firm's strategy cause the firm to seek out external ideas/technology through the open innovation process.

At the mature stage, the firms that are well established also start partnering with startups and fund new small ventures. Hence the average size of the collaborators dips slightly as the firm gets more established.

All these findings are in line with the central idea of open innovation, wherein a firm scans the industry and research organizations to find collaborations to take superior ideas and opportunities to the market, which it cannot do on its own, while exposing its own ideas to the collaborators (Chesbrough 2003). Another finding is that there is a distinct move towards globalization. Even in the case where there are fewer labs or locations (e.g. Holst Center, Lindholmen Park), collaborations from different countries are sought after. In case of more global forums that do not have immediate commercial goals, building standards, or establishing a common model or best practices and framework, seems to be a key driving force, for example, GSMA OneAPI, FOSS 4G, Telecentre, and so on. However, in the case of non-commercial forums that are confined to one or two countries, their main goals seem to include social welfare—to improve entrepreneurship in the region, impart training, help them find funding, and so on. These are also supported by the governments in some cases.

Another interesting observation is that the consortia or forums formed seem inclined to take on many responsibilities across the spectrum—finding ideas, partners, and collaborators in the industry, government, academia, and so on, training, funding startups, finding investors for startups, incubation, helping startups take-off, conducting collaborative events, mass ideation contests, and so on. This again points to the rapidly growing interest and conviction in open innovation. A key takeaway is that open innovation is not new in the industry. Several well-established open innovation forums have their centers running successfully for many years, even decades, for example, Adastral Park in the UK, Lindholmen Park in Sweden, AT&T Foundry in the USA, and so on. This shows that open innovation is a sustainable idea and has rightly attracted many organizations, old and new, into its fold.

Focus is another softer aspect of a forum. An alliance may be considered as more focused if it deals with fewer topics in a sustained manner, and less so if it encompasses many areas (or has a more loosely defined agenda). For example, the forums HiveColab, BongoHive, ActivSpaces, Meltwater, and Miriade have a lesser focus as compared to Cisco EIR, GSMA OneAPI, or KDDI Open Innovation Fund.

By assessing the clusters, it is found that open innovation increases with focus. It also increases by building a strong system for making successful collaborations happen. The collaborations need to include diverse partners such as corporations, government agencies, research institutions, and academic institutions, besides involving individuals and providing strong forums and opportunities for the partners to collaborate. The process takes time and effort to mature, especially if the forum is in a developing country with fewer resources, for example, KINU, KLabs, ActiveSpaces, MEST, and so on, whereas the forums in economically developed countries backed by well-established corporations tend to gain high intensity and productivity fast, for example, Cisco's EIR, KDDI Open Innovation Fund, Verizon's OI center, Telefonica's OI forum.

The Takeaway Box: Managing Collaborations Is Key to Your Success

- Go beyond ordinary collaborations that merely achieve specific business goals.
- Invest in collaborations that promote radical innovations.
- Retain focus of the firm and prefer collaborations over mergers and acquisitions. Investment to an extent in a collaboration is OK.
- Collaborate with academia and research institutes.
- Involve end-users not only as consumers but as co-creators and producers.
- Foster startups; seek out disruptive models yourself.
- Leverage crowdsourcing, open source, and open innovation in all forms.
- You must eventually achieve a good balance between open and closed innovation.
- Contribute to industry standards and government regulations.
- Improve processes and frameworks constantly.
- Go global with collaborations, to the extent you can.

6.7   THE EFFECT OF COMPETITIVE INTENSITY

In this study, competitive intensity in the Telco market of a country was also examined. Competition is closely related to the regulatory environment in that country.

The study (Birudavolu 2015) shows that open innovation dampens the adverse effect of intense competition by turning competitors into potential collaborators (coopetitors). This can further pave the way for cross-holdings, mergers, or acquisitions among the Telcos. Adoption of open innovation strategy redirects part of the competition's force towards helping in improving the firms' own performance, into a compete, co-opt, and cooperate model.

For example, if Telcos implement mobile money and decide to cooperate with each other, the subscribers of one Telco can transfer money to the subscribers to another Telco, thereby increasing the total pool of users, for all the Telcos. Of course, a Telco firm can always offer better rates or discounts for money transfers done within its own subscriber base and incentivize retention. But the Telco firm would still get more revenues overall due to the cooperation with other Telcos. This is because the subscribers would like to have the convenience of transferring money to anyone and may not mind paying a little extra for the convenience, when the other party is a subscriber to a different Telco.

The results of the research in this thesis showed that competitive intensity does not retard the power of open innovation but rather it has a positive effect on the latter. The relationships are summarized in a diagram as follows (Fig. 6.6).

The future competition (Prahalad and Ramaswamy 2004) is all about co-creating value with the customers. The innovation will shift from products and services to experience environments that individuals can interact with to co-create their own compelling personalized experiences. This is key to the organization's future because it is the real source of unique value. All of the organization's resources, including its collaborations, must be deployed towards competing on experiences and co-creating unique value.

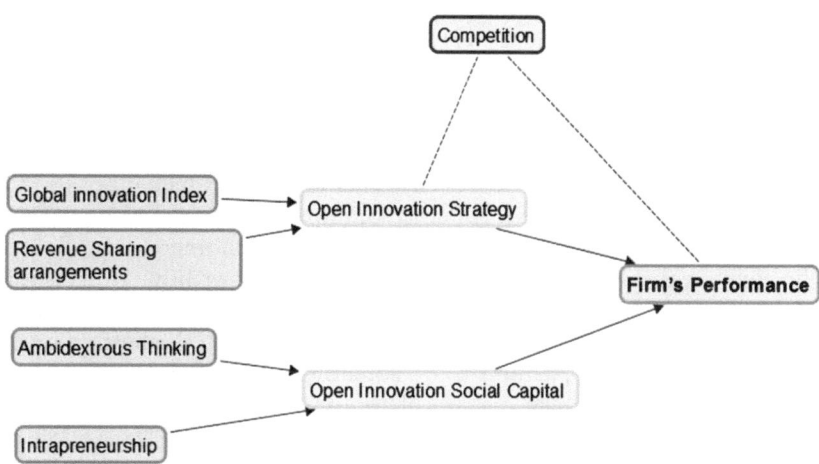

Fig. 6.6 Summary of the basic model (Source: Birudavolu 2015)

In the current day, products and even intellectual property to an extent are considered too static to be of much value to the customers. Open innovation is considered *de rigueur*, that is, "a given". The real differentiator is about how well it is implemented.

The Takeaway Box: Learn How to Use Competition

- Think of competition differently. Their presence will not retard your collaborative power.
- Find ways to collaborate with competitors.
- Examples are building industry standards, frameworks and open-source products, and tapping into competitor's client/subscriber base in specific ways, co-sharing IP development costs, or a third-party partner's costs, or network resources.
- It may help discover new ideas and streams of revenue, improve profitability, and present a strong case to the regulators on issues of major importance to the industry.
- It may help avoid a disastrous price war.

6.8   BUILDING AN OPEN INNOVATION-BASED ICT STRATEGY

6.8.1   Customer Centricity

Open innovation greatly improves customer centricity. Open innovation is inherently customer focused. All collaborations have one goal in mind—maximizing the value and success provided to the customers, regardless of who delivers the products/services to the customer or how. In fact, it is because an organization lacks the capacity to deliver the best possible value to the customers all on its own that it collaborates with the best-of-breed ideas and best-in-class partners to do so, although they're external to the organization and will take a share of the revenues. It is not enough to procure, learn, build, and deploy architectures and technology stacks that are services-centric. The entire thinking and mindset of the organization must become customer-centric. Employees across the rank and file of the Telco must be regularly exposed to customers' and partners' thinking, preferences, priorities, environment, and way of working. There should be a special drive to imbibe this thinking at all levels. Constant training about the Telco's end-to-end ecosystem, value engineering, and value networks will go a long way.

The Takeaway Box: Organize the Entire Firm on the Customer Centricity Principle

- Aim at creating personalized and co-created experience environments with the customers and for the customers.
- Customer centricity is not confined to sales, marketing, or consulting people.
  - It needs to be imbibed across rank and file of the organization, *especially in IT*, and should include every part, such as HR, recruitment, support, and must also extend to partners and suppliers.
- Customer centricity does not mean blindly following what the customer wants right now. Electric light did not come about through continuous improvement of candles.
- The key is to find out what the customer needs, not merely what he wants.
- Use ICT systems to build an ecosystem that is very customer-centric.
- Automate most of the interactions by default, and shield complexity.
- Simplify!

From several points of view, especially customer experience and key stakeholders' experience (especially developer experience).

6.8.2   Building Strategic Capability

Strategy is the "brain" of open innovation. The firm should have a think-tank to study the prevalent open innovation strategies in the industry and build on those strategies. It cannot be left to chance. These strategies cover the full range from revenue-sharing arrangements with partners to forming alliances, mergers, and acquisitions. With the increase in a variety and number of partners, as a natural consequence of open innovation, it is essential for a firm to develop its own frameworks, metrics, and processes related to open innovation.

The firm should build deep intelligence pertaining to customers, partners, potential partners, competitors, agencies, market trends, and technologies, all of which can enable it in taking faster decisions to become nimbler in the market. The most important ability is to evaluate key parameters rapidly (almost automated) for making important decisions pertaining to collaborations and partnerships.

The Takeaway Box: Building Strategic Capability Is Key to Survival, Let Alone Become Greatly Successful

- Building strategic capability is not the same as having a good strategy.
- A firm must build and maintain a practice for continuously scanning the landscape for potential collaborations, disruptive business models, technological innovations, competitive moves, regulatory changes, and so on.
- This practice must essentially take input from every part of the organization. There could be subtle changes occurring in the ecosystem that are observed only by some corner of the organization. It must not be left to accident, chance or casual meetings, or processes meant for other purposes.
- This is above and beyond the usual market intelligence. For example, the engineering and recruitment departments together may decide to start a collaboration with an academic institute to further research in a certain area, and to fund PhDs in that area. They could prototype a kind of service which can be tried out by plugging into the firm's platform through an experimental sandbox.
- Measuring every aspect of the capability is key to building strategic capability, for example, knowing cloud economics may be essential for shortlisting vendors. This capability should be built well before it is needed.

6.8.3   The Extended Model

While in the basic model we consider the collaborators to be those partners who directly provide services by plugging into the firm's ecosystem, in the extended model we include all kinds of collaborators, namely, government organizations, academic institutes, financial institutions, end-users, and even competitors. The diagram for the extended model is shown in the next page. In the depiction, for simplicity's sake, all these are summarily shown to play through the country's Global Innovation Index, but in practice many of them may directly work with the firm to varying degrees. To prevent cluttering the diagram with too many arrows, only the arrows acting through the GII are shown (Fig. 6.7).

The Takeaway Box: Ultimately Your Ecosystem Is Your Only Real, Sustaining Strength

- Nurture and grow your ecosystem, not merely your organization.
- Continuous incremental improvement through collaboration with the ecosystem will make your organization antifragile, that is, less susceptible to "black swan events", which are defined as sudden, unexpected cataclysmic events.
- This is because some parts of the ecosystem that you are collaborating with will have the disruptive capability for good growth into new markets, even if some of the existing models wither away. This is happening with Telcos where their older staple of voice revenue has all but vanished, but there are opportunities to flourish in other areas.
- Thorough requirements engineering and market research must precede the undertaking of building a platform and ecosystem.
- Recognize that good ideas can come from anywhere, such as from a student intern coming from an academic institute. Aggressively document and reward:
    ideas, innovation, and collaborations—both within and outside the organization.
- Your survival and growth in the future may depend on a small idea that you've nurtured carefully today!
- Develop a sense of urgency in evolving the ecosystem. Avoid analysis paralysis.

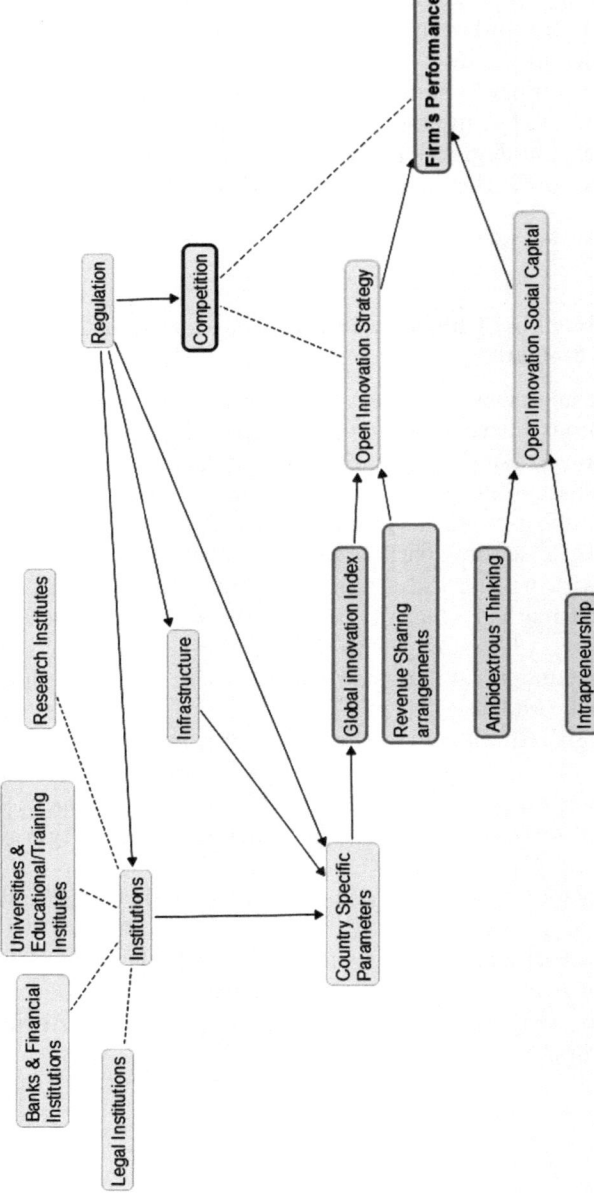

Fig. 6.7 Depiction of the extended model (From the author's PhD research, Birudavolu (2015))

6.9 THE TRACTION GAP FRAMEWORK

As all companies need to think like startups and launch new innovative products or innovate on existing products, it is imperative to adopt a framework that will help in implementing the strategy successfully. TGF or the traction gap framework (Wildcat Ventures 2017) is a good tool to use. Venture capitalists use this to evaluate startups and help them succeed. The framework needs to be tailored for each individual startup.

While the TGF is aimed at startups, all companies would do well to imbibe it, relative to the specific innovation being launched. Briefly, the TGF defines five milestones along the path to viability:

1. *MVC*—the minimum viable category, which is a clear definition of the category in which the product is being launched, and the strategic reasons why this category is worth pursuing, that is, there is a good chance of success.

2. *IPR*—initial product release, which is the first, initial release of the product, that is, general availability of the product to the public. This release is used to bootstrap the future releases by seeking feedback from the market. This milestone is the entry point into the traction gap framework.

3. *MVP*—minimum viable product, which is the version of the product that has all the minimum essential features required to make the product survive/float in the market, as measured by the customer validation metrics.

4. *MVR*—minimum viable repeatability, which is the milestone in which a strong, flesh-and-blood version of product is released into the market. This version demonstrates a good product/market fit and shows business model feasibility by providing a proper solution to the customers.

5. *MVT*—minimum viable traction is a milestone reached after the MVR has shown several quarters of growth with progressive improvements in the product. With this the product may be deemed as ready to exit from the TGF. This milestone signifies that the product has accomplished good traction in the market. Now it needs to scale over the next five or six quarters.

The four core architectural pillars described by TGF are product, revenue, team, and systems. TGF considers these as the core building blocks of any company, regardless of the organization's size or maturity, and uses these to evaluate the startup and improve chances of the product's market success.

6.10   THE INNOVATION MATRIX: WHICH INNOVATION INITIATIVE TO CHOOSE?

To help an organization define an innovation strategy that fits the company, the Board of Innovation has defined an innovation matrix (Pirenne 2017):

This matrix helps guide an organization in choosing the type of innovation initiatives that will best suit it. The two axes are capabilities and commitment (which is intensity).

Commitment indicates the investment that a company is willing to commit to an innovation program, which ranges from a small workshop or an innovation challenge to a full-fledged center of excellence, as indicated in the matrix.

Capabilities indicate the locus of innovation, ranging from reliance on internal resources to external parties for making innovation happen. The nine combinations are discussed in detail in (Pirenne 2017), as well as how to use the innovation matrix.

It is strongly recommended that over a period an organization should evolve and develop both internal and external innovation capabilities; hence it should have both a center of excellence and a startup fund. And as the size of the organization grows, it should continue to run both structured and non-structured programs depending on the problem in question. One size will not suit all situations. However, to get started and progress along the path of innovation, an organization may choose a path through the nine cells of the matrix as it deems fit.

6.11   CONCLUSIONS

No organization however large or powerful can meet all its customer needs solely through creating products/services by dint of its internal resources only. Organizations need to open their business/innovation models in a strategically planned and proactive manner. The whole idea is for the organization to create a best-of-breed integrated solution portfolio for the customers (thereby matching or exceeding demand-side) by collaborating with the specialized products/services that the market offers (from the supply-side). Strategically, the organization should become an integrator that combines its own offerings and expertise with those of its partners, to provide cutting-edge and innovative solutions to the current and unmet emerging needs of the customer. The organization's chief differentiator over time are the historical and live data generated from the operations.

The framework and the components of strategy for innovation are discussed in this chapter.

The Takeaway Box: Innovation Strategy Is a Given, Execution Is the Differentiator

- If you don't even have an innovation strategy in the first place, you don't stand a chance in today's market.
- Strategy should aim at creating superior value through an innovative combination of business, technology, and design, achieved through every means (internal resources and collaborations).
- Strategy should result in creating a collaborative, flexible, and customer-centric organization.
- A great execution of strategy will yield superior customer satisfaction, operational excellence, and profitable growth.
- Setting up a good rewards system (e.g. revenue-sharing arrangements) is an important part of a good open innovation strategy.
- Use a new lens that looks upon the following as key assets for the organization: customers, users, employees, partners, academia, regulators, and even competitors. All of them serve us with market information, shape our strategy, and provide us with collaboration opportunities. Consider all of them while formulating strategy, the first four being the most important.

REFERENCES

Afuah, Allan, Innovation Management: Strategies, Implementation and Profits, 2003, Oxford University Press, ISBN: 0195142306, 9781621984726, 1621984729.

AirTel Open Network. (2016). Retrieved from www.airtel.in/opennetwork.

AirTel Samsung Collaboration. (2018). Retrieved from www.airtel.in/press-release/01-2018/airtel-and-samsung-join-hands-to-transform-india-into-a-smartphone-nation.

Bigliardi, B., Dormio, A. I., & Galati, F. (2012). The Adoption of Open Innovation within the Telecommunication Industry. *European Journal of Innovation Management, 15*(1), 27–54.

Birudavolu, S. (2015). *Open Innovation in ICT: An Empirical Assessment of Global Telecommunications Services.* Unpublished thesis submitted for PhD at Indian Institute of Foreign Trade, New Delhi.

Birudavolu, S., & Nag, B. (2015a, February). Attributes of Open Innovation ICT Forums, *International Journal in Management and Social Science, International Journals of Multi-Dimensional Research.* Retrieved from http://ijmr.net.in.

Birudavolu, S., & Nag, B. (2015b, March). The Digital Services Landscape: An Empirical Study of Open Innovation ICT Forums, *International Journal of Multidisciplinary Research.* Retrieved from http://www.ijmra.us.

Birudavolu, S., Nag, B., & Wali, O. (2016). Open Innovation in ICT Services for Quadruple Helix Model. *Proceedings of the International Conference on Internet of things and Cloud Computing – ICC '16, 2016, Article 33,* ACM Conference, Cambridge, United Kingdom.

Chesbrough, H. (2003). *Open Innovation: The New Imperative for Creating and Profiting from Technology.* Boston, MA: Harvard Business School Press.

Chesbrough, H. (2006). *Open Business Models: How to Thrive in the New Innovation Landscape.* Boston, MA: Harvard Business School Press.

Christensen, C. (1997). *The Innovator's Dilemma: When New Technologies Cause Great Firms to Fail.* Harvard Business School Press.

Christensen, C. M., & Raynor, M. (2003, October). *The Innovator's Solution: Creating and Sustaining Successful Growth.* Boston, MA: Harvard Business School Press.

Dahlander, L., & Gann, D. M. (2010, July). How Open Is Innovation? *Research Policy, 39*(6), 699–709. https://doi.org/10.1016/j.respol.2010.01.013.

Elmquist, M., Fredberg, T., & Ollila, S. (2009). Exploring the Field of Open Innovation. *European Journal of Innovation Management, 12*(3), 326–345.

Erkens, M., Wosch, S., Luttgens, D., and Piller, F. (2013). *Measuring Open Innovation – 3 Key Principles to Improve Your Innovation Measurement Practices.* Retrieved April 15, 2015, from www.innovationmanagement.se/2013/06/14/measuring-open-innovation-3-key-principles-to-improve-your-innovation-measurement-practices-part-1/.

Firican, G. (2017, February 8). The 10 V's of Big Data, TDWI—Transforming Data With Intelligence. Retrieved from https://tdwi.org/articles/2017/02/08/10-vs-of-big-data.aspx.

Intelligencia Limited. (2018). Retrieved from https://www.slideshare.net/IntelligenciaAnalytics/intelligencia-limited-brochure.

Johnson, G., & Kevin, S. (2002). *Exploring Corporate Strategy: Text and Cases* (6th ed.). Prentice Hall of India.

KDDI Financial Reports. (2018). Retrieved from http://news.kddi.com/kddi/corporate/english/ir-news/2018/01/31/pdf/kddi_180131_e_main_3eC6qS.pdf.

LUCY. (2018). *Sustainable Environmental Monitoring Platform*. Retrieved from www.thinkphi.com/lucy.

NDTV. (2018, February 27). To Bring In-Flight Data Connectivity, Airtel Joins Global Consortium. Retrieved from https://www.ndtv.com/business/to-bring-in-flight-data-connectivity-airtel-joins-global-consortium-1817523.

Nesse, P. J. (2009). *Open Service Innovation in Telecom Industry—Case Study of Partnership Models Enabling 3rd Party Development of Novel Mobile Services*. Telenor Research & Innovation, Service Innovation Group.

OpenCellular Project. (n.d.). Retrieved from https://code.facebook.com/posts/1754757044806180/introducing-opencellular-an-open-source-wireless-access-platform.

OpenCompute Project. (n.d.). Retrieved from www.opencompute.org.

Pelz, D. C. (1978). Some Expanded Perspectives on Use of Social Science in Public Policy. In M. Yinger & S. Cutler (Eds.), *Major Social Issues: A Multidisciplinary View* (pp. 346–357). New York: Free Press.

Pirenne, V. (2017, July 25). *The Innovation Matrix – A tool to define the Innovation Strategy that best fits your company, The Board of Innovation*. Retrieved from https://www.boardofinnovation.com/blog/2017/07/25/innovation-matrix-tool-to-define-your-innovation-strategy/.

Prahalad, C. K. and Ramaswamy, V. (2004). *The Future of Competition: Co-Creating Unique Value With Customers*. Harvard Business School Press, ISBN-10: 1578519535; ISBN-13: 978-1578519538.

Ramaswamy, V., & Gouillart, F. (2010). *The Power of Co-Creation*. Simon and Schuster, ISBN: 9781439181041.

Telecom Drive, July 10 2017, Digital Transformation: Airtel Presents 'Project Next' – its Digital Innovation Program, http://telecomdrive.com/digital-transformation-airtel-presents-project-next-digital-innovation-program/.

Telecom Infrastructure Project. (n.d.). Retrieved from www.tip.org.

WEF. (2017). *Digital Transformation Initiative Telecommunications Industry Whitepaper, World Economic Forum*. Retrieved from https://www.accenture.com/t20170411T115809Z__w__/us-en/_acnmedia/Accenture/Conversion-Assets/WEF/PDF/Accenture-Telecommunications-Industry.pdf.

West, J., & Gallagher, S. (2006). Patterns of Open Innovation in Open Source Software. In H. Chesbrough, W. Vanhaverbeke, & J. West (Eds.), *Open Innovation: Researching a New Paradigm* (pp. 82–106). Oxford: Oxford University Press.

Wildcat Ventures. (2017). *The Traction Gap Framework*. Retrieved from http://wildcat.vc/wp-content/uploads/2017/09/Traction-Gap-Framework-9.14.17.pdf.

Yahoo Finance, Razer. (2018). Retrieved from https://finance.yahoo.com/quote/1337.HK?p=1337.HK.

## CHAPTER 7

Social Capital: The Innovation Culture

Culture eats strategy for breakfast, lunch and dinner.
—Peter Drucker

7.1 INTRODUCTION

What is social capital? We can use these working definitions:

- "Networks together with shared norms, values and understandings that facilitate co-operation within or among groups".—*OECD*
- "Social capital is a form of economic and cultural capital in which social networks are central, transactions are marked by reciprocity, trust, and cooperation, and market agents produce goods and services not mainly for themselves, but for a common good".—*Wikipedia*

Innovation social capital (hereafter called social capital) of a firm refers to the innovation culture fostered within the organization. This is ultimately reflected in the innovative products and services that the firm offers. The social capital is the innovation capability of the human resources in the firm. While innovation strategy (hereafter called strategy) is the brain, the social capital is the heart of innovation. A firm must consciously nurture a culture of open innovation in the organization. An innovation strategy is only as good as its execution. And the execution is carried out by the organization. It cannot be done merely by constituting a think tank at the top or by the heroic solo efforts of a few charged-up individuals

© The Author(s) 2019
S. Birudavolu, B. Nag, *Business Innovation and ICT Strategies*,
https://Doi.org/10.1007/978-981-13-1675-3_7

somewhere deep in the organization (Hamel and Prahalad 1996). A culture of innovation must be imbibed across the rank and file of an organization, starting from the very top and spanning all the way to the lowest echelons of the organization. But how do we go about doing it?

This chapter deals with this question of how to build innovation social capital, especially that of open innovation in any organization, by leveraging ICT.

7.2   MEASURING INNOVATION SOCIAL CAPITAL

The social capital can only be measured indirectly, for it to carry any objectivity or a uniform yardstick across firms or divisions within a firm. In the final analysis, one must measure the innovation social capital against the innovation content in the actual products and services launched by firm in any given period. Otherwise the social capital metric cannot be deemed to be effective.

There are two important measures for social capital in the firm: ambidextrous thinking and intrapreneurship. These useful parameters are well regarded in open innovation literature (Fasnacht 2009).

7.3   AMBIDEXTROUS THINKING

Ambidextrous thinking capability is one of the variables that measure the organization-specific social capital for innovation. The products and services launched by the firm can be analyzed for their ambidextrous thinking quotient.

Essentially, ambidextrous thinking involves striking a balance (and even synergy) between opposite, such as:

- Closed vs. open innovation
- Simplicity vs. richness of functionality
- Exploitative vs. exploratory innovation
- Efficiency vs. flexibility
- Product focus vs. client focus
- Depth vs. width (i.e. niche vs. mass market)
- Proprietary and central systems vs. flexible service-oriented architectures
- Production vs. consumption of services and content (e.g., in social networks the end-users are "prosumers)

Ambidextrous thinking is reflected in the uniqueness, richness, and the manner in which the services are launched, reconciling the trade-offs between opposing forces. These are measured along the dimensions mentioned above, using the ratings. For software, for example, there could be several pairs of contending factors:

1. *Usability vs. Security:* A higher level of security with multiple access controls could result in increasing pain of use, due to all the severe procedures and restrictions (resulting in lower usability or bad user experience). Conversely a greatly easy-to-use interface with no restrictions could compromise security severely. For example, to make a system very secure, if we make a user do all of the following: enter password with many rules, force biometrics (fingerprint/retina scan), insist on an OTP (one-time password) sent to his mobile number or email account, and to top it all, also keep stringent time-out locks on the systems, then it becomes suffocating for the user. On the other hand, to make it very easy for the user, if we allow free access with only one minimal strength password, then it may be a huge security hazard, which could compromise all the systems.

2. *Cost vs. Performance or Cost vs. Scalability:* Frequently, performance (or scalability) comes at a price. For example, for a retail online website, it is easy to recommend that the organization should buy capacity that caters to the projected peak load during a certain upcoming festival date. However, the peak load of a festival date could be *several times the peak load* of a normal day. It does not justify sinking in huge capital investment that will be put to use, only for a miniscule fraction of the time, or alternately even signing an abnormally large contract with a cloud vendor to provide for the high peak capacity. A much better alternative could be to size the system at normal capacity and put measures in place to rapidly scale-up and ramp-down the capacity. This can be done in a way that minimizes the risk window.

3. *Functionality vs. Footprint:* A smaller, easy-to-deploy/load software may have to cut down on functionality to keep the size small and achieve nimbleness.

It is important for a product/service to choose the right pairs of contending factors to balance, while allowing imbalance in other pairs. Conversely, one may deliberately choose imbalance in a few pairs of factors, while balancing the other pairs. For example, the choice could be to keep running costs as low as possible, for example, running fuel costs per kilometer for a budget car. This one decision would automatically determine several other trade-offs, such as weight of the car or luxury features. Thus, the steps in ambidextrous thinking involve:

1. First, reduce the essence of the concerned matter/area to a handful of orthogonal pairs of parameters, each of them representing a specific trade-off. Identify as many pairs of factors as possible. In fact, get these from research already done in the concerned area
   (a) This exercise will result in pairs such as performance vs. cost, stability vs. efficiency, size vs. mobility, and so on.
   (b) Focus on the critical pairs first.
   (c) Determine trade-offs for each pair.
2. Then find the scale for each parameter; get it from already known research.
3. Mix and match parameters, by varying the degree on the scale—identify with existing entities in landscape.
4. Take decisions on which ones to focus on, for a particular product or service.
5. Craft a unique product by adjusting the trade-offs on the remaining pairs.
6. Find a new combo/niche that doesn't yet exist in the landscape (use "pain" factors to guide practicality).
7. Check if new combo will click.
8. Find out what it takes to make the new combo work (e.g., by introducing new components, etc.).
9. Constantly try to add new orthogonal factors, it may launch you into a new realm altogether.
10. Craft a unique product by adjusting the trade-offs on the remaining pairs.

Adjusting and tuning the trade-offs on the different pairs can result in various unique, innovative combinations, bringing about products and services that one has never thought of before. This is the key to innovation.

7.4 Intrapreneurship

Intrapreneurial thinking is "market savviness" and refers to both the aggressiveness with which ideas are pushed into the market, as also a fine sense of alignment to the market. The chapter on focus delves into how an organization should develop focus and how to launch focused products that meet the market needs. In a startup driven by an entrepreneur, the culture is very straightforward. However, in larger companies, the culture needs to be fostered among the employees. All that is discussed in the chapter on focus holds here as well. The limitation here is that every employee in every function does not do market research or establish focus for the product or company. Then how does one imbibe intrapreneurship in all other functions and even with routine operations or incremental feature releases?

Intrapreneurship is a softer aspect in a firm and is cultural in nature, but it has great transformative power.

Intrapreneurship is firm-specific (Fasnacht 2009), and in a firm, it should be measured and tracked. The considerations for rating intrapreneurial attitude are as follows:

- The market harmonization/penetration in the desired subscriber market
- Whether the launched service extends/complements the existing services into the current subscriber base or ventures into entirely new markets
- Whether the launched service reuses and builds upon existing services
- Width (relative to the size of the subscriber base targeted—this could be existing or new subscribers)

Thus, intrapreneurial thinking is all about:

- Matching ideas with market need, aligning them well to the market
- The aggressiveness with which the service is pushed into the market; selling the idea
- Tweaking environment, making the change happen, finding the means to realize it
- Being proactive, building partnerships

Feedback is an important aspect of intrapreneurship. In customer facing roles, every interaction is an opportunity. In operations, the data tells a rich story.

Traditionally, the teams in marketing, sales, and product development handle these, but in a customer-centric organization that is highly innovative and flexible, every single role, function, and process need to imbibe these. Good ideas can come from anywhere, within or without the organization, and everyone in the organization holds the responsibility for market alignment of the ideas that they bring to the table. The innovation processes, tools, and knowledge bases will help individuals and teams in creating and honing their innovations.

Intrapreneurship is an important part of innovation social capital. It is pointless to be great at ambidextrous thinking and generate unique ideas, if the ideas do not target the market properly, or be unable to push the ideas through because one is unable to sell the idea, or find the means, or fails to identify and build the right alliances, or is incapable of making the necessary changes internally and externally. Intrapreneurship enables one to explore all possible means to find pathways to the market. Even in generating ideas, an intrapreneur considers extending or repurposing and reusing existing products and services, for new versions of the existing services or for new purposes.

It is great to sell an experience to the customer, rather than a product, solution, or a service. If it resonates emotionally with the customer, then the product can command a premium value.

A fresh look at the status quo really helps. One can do this in a variety of ways, namely, cross-reviews by other teams, hiring outside experts and consultants, joint projects with different teams, external partnerships and collaborations, comparison studies of how a thing is accomplished in other organizations and geographies, sponsoring studies and projects in research institutes, and digging into history for what led to the current state, especially about what assumptions were made in the past, what were the constraints in the past (which no longer hold), and so on.

It is great to carry a consultative mindset for developing both ambidextrous thinking and intrapreneurship. The organization must hold workshops and training programs in these two areas.

The Takeaway Box: Neglect Innovation Social Capital at Your Own Peril!

- It is impossible to realize innovation strategy in an organization without building a critical mass of social capital first.
- Intrapreneurship and ambidextrous thinking are the twin pillars of social capital.
- It is essential to create and sell an experience to the customer rather than product, solution, or service.
- Both can be systematically taught and imbibed in an organization.
- Social capital becomes self-propelling and rewarding when practiced well (into the organization's DNA).
- An organization should strengthen and improve both ambidextrous thinking and intrapreneurship through knowledge bases, training programs, pilot projects, prototyping, mentorship, research, and collaborations.

7.5   Constant Awareness of Resources' Capability and Utilization Is Key

It is estimated that the USA alone wasted over $30 billion on unused software over the course of a 4-year study, and the global average came to about 37% wastage per company (Miller 2016). With business models, technology, and processes shifting very rapidly, resources may become redundant fast. This not only includes resources for storage, computing, networks, but also people and their skills and experience. Entire projects may be shelved suddenly due to changes in the market, due to mergers and acquisitions, or even due to technological and business obsolescence. There could be organization restructuring, downsizing/right-sizing. People may leave and new employees may join at different points in time. It is vital for an organization retain its organizational culture and capability amidst all these changes.

It is also necessary to be very flexible to redeploy resources in entirely new ways. For example, the dot-com bust of 2001 left in its wake staggering piles of unused hardware from companies that were closed or downsized. This unutilized hardware was innovatively used through pooling, and remotely renting out via the Internet and the computing and storage in smaller units on a usage basis, thereby spurring the growth of data centers and cloud computing.

While it is important to build architectures and stacks that are very flexible, for example, cloud, platform, virtualization, microservices, and so on, it is even more important to build a flexible mindset in an organization. All the skills, talents, and resources in a changing environment should carry a guiding principle of maximizing resource utilization on a value basis, that is, applying resources to high yield areas. The utilization itself should be measured (in dollar terms and other appropriate measures), as to how many resources are being used to support current operations and existing business models versus investment in future, budding growth areas of high promise. This cannot be left to chance.

Whatever be the vision, intent, strategy, or execution, the actual utilization shows the correct picture of resource deployment. There should be conscious checks and balances. The social capital must include a mindset of high resource utilization and an abhorrence of wastage. These should be listed in the key result areas of performance goals set for people across the rank and file of an organization. The "Reduce-Reuse-Recycle" mantra is not merely good for sustainability and a greener earth but may help in the survival of an organization.

All innovations that improve resource utilization in new productive ways should be immediately encouraged. ICT/Telecom companies build a lot of capacity end-to-end, with growth in mind, to address peak traffic and meet SLAs even in contingencies. Sadly, this also hits their utilization and ROI—the Capex and Opex to run this surplus capacity.

People in the organization must constantly find innovative ways of putting this buffer capacity to better use, while carrying minimal risk of jeopardizing existing commitments. Procurement of new resources should be done as an exception, needing much justification, and only after assessing the current capability thoroughly, against the vision and strategy of the organization. Alternately, one can sign contracts with vendors, such as cloud providers, stipulating appropriate SLAs in order to accommodate peak utilization. For doing this, one can share prediction patterns and historical data with them, especially emphasizing seasonal variations.

The Takeaway Box: Constantly Keeping Resource Utilization in Mind Can Spur Innovation

- A capability framework in an enterprise can help assess what the enterprise is capable of accomplishing with its current resources. In short, it is the wherewithal of an organization. The capability should be constantly measured vis-à-vis the goals.
- In a constantly changing world, innovation helps put the resources to better use, either by reducing wastage and improving the bottom line or by launching new innovative services that add revenue streams.
- A resource utilization mindset adopted across the rank and file of an organization greatly helps the balance sheet of a company. The measures should be embedded in vision, planning, processes, operations, systems, and innovation.
- Innovation processes and measures should include the specific aims of reducing waste and improving utilization, reuse, and procurement as an exception.
- Innovation for improving resource utilization should not be confined to the organizational resources; they should be applied to the industry and its problems. This will yield new business models, services, and revenue streams, especially when done in collaboration with partners.

7.6   LEARN HOW TO GENERATE AND USE DATA

The biggest generational shift in culture in organizations is that of usage of data. This is doubtless propelled by the rise of ICT systems and automation. Unfortunately, the social capital has not at all kept pace with the technology. Most organizations are laggards in knowing how to really use data. Minimum compliance to minimal standards seems to be the norm everywhere. While building the right architecture and tools are described in other parts of this book, this section deals with the question of how to imbibe the culture of using data.

The moment an IT/ICT system is built, deployed, and switched on in production, it starts spewing out a ton of data every hour. For instance, if a bank commissions a banking portal, millions of people start transacting on it round the clock and instantly generate an astronomical number of records pertaining to login, inquiries about their accounts, fund transfers, change of personal details, and so on. Organizations and people increasingly rely on automation and information processing to help service their customers. To achieve both depth and scale, they find themselves inundated with data that only a few have the capability to process, use and benefits from.

Data tells a story that is important to extract. The story itself is shaped primarily by a handful of key factors, such as:

1. The underlying model on which the system has been built, along with all the assumptions made (theoretical and practical), including dependence on other systems
2. The way the system has been configured and customized
3. The varying circumstances and the environment in which the system is operating
4. The way the system is being used by the people and how data is being generated
5. The way the data is being processed and interpreted

Each of these interventions needs expertise and social capital in the organization to comprehend and exploit the data. While an organization can rely on external data science expertise and alliances to help them, it is essential to build a critical mass of "data literacy" across the organization for running the business, defining and developing systems, and planning the next steps. Data is the lifeblood of the modern organization and it is ill-advised to depend completely on external parties for understanding the data of one's own business.

For example, every team and system in the organization must know its boundaries and limits as precisely as possible in terms of metrics, and each department and team must understand their own respective boundaries. The limits at a higher level in the organization should automatically translate to all the limits at the lower levels. Examples of limits could be budgets, or operational parameters such as permissible loan limits that can be given out, or access limits to resources or information. If an exception is flagged at a higher level, it should be traceable as to which all systems or departments have been impacted or perhaps where the exception originated from. The exception could be a violation of regulatory or organizational norm, or an outlier event, such as a sudden transaction of an abnormally large amount, out of the norm that has been flagged by the

AI/ML system. The teams right from operational to security/compliance and the management should know how to interpret it and what action to take.

Insofar as innovation is concerned, data is a crucially valuable source of business ideas and for driving deeper customer engagement, as well as for finding new customers. Data monetization is a major component of open innovation (Birudavolu 2015). The data generated by an organization is extremely valuable for another organization, which can partner together to produce different business models. There is in fact a supply chain of data that is used by many e-commerce retail companies. For example, the browser cookies, which capture user preferences and information about websites that the user has visited, are much sought after by many media and marketing companies who sell them to companies like Amazon and Google, who in turn use them to target the user with specific advertising.

Information is invaluable and the sooner every employee learns this fact, the better.

The Takeaway Box: Data Is the Life-Blood of the Modern Organization: Data Literacy Is Important!

- Google, Microsoft, Facebook, and others spend billions of dollars every year, running millions of servers worldwide to give away a host of *completely free services* for web search, mail, calendar, social media including video chat, storing documents, and so on. But why…? Are they doing charity?

  The reason is no secret at all—it is because the data they get from the users and businesses *is the basis* for both innovation and running their business. Your data is *that* invaluable!
- There needs to be not only a data strategy for the organization but also a culture of using data.
- Social capital must include data literacy and expertise.
- Realize that there is a flood of data that cannot be handled by humans. It needs ICT tools, right from basic to advanced levels of sophistication.
- If there is inadequate data, then this is a problem that needs to be fixed immediately!
- Build templates and tools for manipulating and using data specific to your organizational needs. Every employee at every level needs

(*continued*)

(continued)

to have good sense of what the data needs are and how to use the data.

- Everyone in the organization needs training on the tools appropriate to his role to produce and process data.
- If employees are neglecting data due to lack of culture or capability, then this is the easiest way to kill the organization in today's world.
- Data can be used in various ways—running operations, data mining for insights, innovation, taking decisions, forming alliances, and even for monetizing.
- As there are no intellectual property rights or patents for data exactly equivalent to that for ideas and innovations, the organization needs to be extremely careful about data security and privacy. Regulations mandate security and privacy only to an extent. For example, there is no rule that a hospital needs to share the *complete* information they have on a patient with either the patient or with another hospital that the patient visits after a few years, or with a medical insurance company.

7.7   A Profit Center Mindset; Not a Cost Center

Even while experimenting and innovating, it is essential to instill the P&L mindset in the organization. The best way to do this is to find ways to estimate/measure both costs and value, thereby one can track value improvements and cost implications at every stage. This holds true for non-profit organizations also. While their goal may not be monetary in nature, but it most certainly is aligned to value creation of some sort. If the work involves development of concepts, whose implementation is still far removed, then there should be still some notion of value and cost, even guidelines and thumb rules to guide the teams. The focus should be on how to link cost with improving productivity and the importance of small and big innovations to drive it.

Consultative thinking helps a lot, as it is all about problem solving and identifying and creating value. All the key members of the organization should undergo training and coaching in consultative thinking, regardless of whether they are building products or are in the engineering division.

7.8   SECURING YOUR INNOVATIONS

Innovation and security always go together in two ways (and for two reasons):

- Legal: For intellectual property protection, in terms of patents, trademarks, documentation, contracts, and so on.
- Technical: To keep track of the creation, classification, metamorphosis, and usage and to prevent theft. This includes the innovation process, knowledge bases, and cybersecurity.

A high-profile example is Qualcomm, which has got into several lawsuits related to intellectual property and unfair patents, worth billions of dollars with Apple, the European Union, as well as other regulators in China, Taiwan, and South Korea. Its takeover of the Dutch chipmaker NXP (Sun 2017; Looper 2017), the company's valuation, and future were at stake. The events could have affected both NXP and Qualcomm's prospects as there could be large liabilities that the acquirer will inherit. The bid failed after Qualcomm could not succeed in getting regulatory approval from China (Martina and Nellis 2018).

As technology, intellectual property, and partnerships become the lynchpins for wealth creation, the disputes and litigations are bound to increase. Businesses are at risk from both property infringements and theft by outsiders. Conversely, the management of every organization needs to be careful that on its part it shouldn't engage in such a malpractice, knowingly or unknowingly. These things can happen deep down in the organization without the management's knowledge or consent. And the organization should put strong measures in place to prevent any such occurrence.

As a culture, organizations need to thoroughly document their ideas and innovations through all the stages, or they cannot use or harvest them when needed. With different organizations collaborating to share information, there should be clear definitions and demarcations on who owns what and also on what a party is permitted to do with the shared data and knowledge owned by another organization. Without having these measures in place, the entire structure of open innovation will collapse. The organization will not be able to monetize or capitalize on its innovations. Legal and technical protection should be factored in, right from the beginning, for all the intellectual property being created from innovation in the organization. It should not be done as an afterthought. Legal cases are expensive, and it is very difficult to prove rights if there are disputes.

There is an ever-looming threat of cyber espionage and leakage. A plethora of technologies and hacking skills are easily available for leaking and stealing information on a large scale; hence, security has assumed an extremely a vital role in the social capital of an organization, in a way that is unprecedented in history. Naturally, cybersecurity is foremost on the agenda of all nations and organizations.

Security is, first and foremost, a mindset and a culture and not just a set of tools and checklists. Hence, it needs to be imbibed in the social capital, not merely drilled in mindlessly through procedures. In fact, major security hazards occur due to overbearing processes that are difficult to follow. Most of the security measures deployed should be non-intrusive and invisible, but powerful and effective.

The Internet is the fundamental driver of the entire ICT industry. However, the Internet itself was never designed to handle this scale and complexity, like carrying entire businesses online on a global scale. There are many problems inherent in the architecture and implementation of the internet. Attempts to adapt the Internet to modern needs, such as adding IPv6 to increase the number of IP addresses, or increasing the bandwidth, computing, and storage to cater to the increasing volume and load, do not fundamentally change the underlying architecture and design of the Internet. These loopholes and flaws present real dangers, notable among them being the fragility of security. This is the preeminent reason why providing security is such a big business in the industry today. It is because both the web and the underlying Internet are inherently very insecure. There are deeper plans underway, such as ICN (information-centric networking) to change the fundamental architecture of the internet, for IoT, superior security, performance, and so on. However, these will need time to take shape.

It is all too imperative then that security concerns become a part and parcel of the *psyche* of the organization. Leave alone running businesses and websites online, even simple online activities such as email or web browsing are fraught with grave dangers, such as identity thefts, ransomware, denial of service attacks, wiping out of bank accounts, and corporate espionage. The social capital must include an acute awareness of the risks in IT/ICT.

Tools, technologies, and checklists can only go so far. It is the mindset or the sixth sense that people need to adopt towards security. Let's take an analogy to drive this point home. People should sense danger all around, much like a lady in the following situation might do:

Consider a tough, crime-ridden, sprawling city that has drug cartels ruling the town, gang wars going on, and groups of street thugs roaming around. Mugging is rampant, and there is practically no law enforcement.

Now think of a young, attractive lady, who is new to the town. She is wearing expensive jewelry and clothing and finds herself accidentally walking in a dark alley at night in this city.

You are like the lady in the above example, the crime-ridden city being the internet!

This analogy to the Internet is no exaggeration. This is the exact situation you and your company are in now, in a virtual sense. Cybersecurity is paramount. Employ every means necessary to thwart off present and future dangers.

Classify all digital assets as per C-R-O-T (critical, redundant, obsolete, and trivial). First, people need to be aware of the different categories of digital assets. The policies, systems, and training follow next. The critical assets are the "crown jewels" of the organization and always constitute a tiny fraction of the total assets (typically 5%), and their security, accounting, and lifecycle management are of paramount importance. The best way to do this is to identify and isolate these and hold them on the cloud with fortress-like security. Redundant digital assets are those that are duplicated in many places, such as in different systems or mailboxes. For example, a document could have been mailed to a group and lives on in different mailboxes, folders, and user devices across the organization. Redundancies should be bulk managed efficiently in an automated manner to constantly remove duplicates. As far as possible, only pointers to digital assets should be circulated. All digital assets should carry an expiry date, beyond which they can be removed. Obsolete assets are like documents that have been archived but have lived beyond the legally stipulated timelines. There is a tendency to ignore these, as these are forgotten and usually not used. But these may pose great risks. These obsolete assets are best eliminated in a timely manner, because they could contain legal, HR, or accounting data, which, if stolen or exposed to the public, could rake up all sorts of old and new issues much to the embarrassment and reputation of the company. Trivial assets, like personal photos and unimportant files, unfortunately take up bulk of the space (even 50%) and need constant and severe curtailing. They need to be purged regularly. People should also be trained to strictly separate personal assets from those of the organization.

An Example:

The company Sony Pictures was hacked in 2014 (Sony Pictures Hack 2014), and a lot of embarrassing material from their servers was put out in the public. This included personal information of employees and families, details of executive salaries, and email communication between employees. There was an un-estimated business loss as the hackers released unreleased pictures' material into the open. Also, many servers were erased of information by the hackers. It was also a major PR disaster for Sony Pictures.

The Takeaway Box: Better Be Safe than Slaughtered!

- Unorganized ideas and innovation are almost useless! Most of the ideas do not even make it past the desk of the innovator. Build processes to capture the ideas, store, organize, and classify them extremely well for retrieval and use.
- Need to protect your intellectual property. Find ways to measure the ideas to pick the ones to patent or trademark.
- As per Robert S. Mueller, Director of FBI:

  "There are only two types of companies: Those that have been hacked and those that will be hacked."

- And peculiarly, the first type that have been hacked consists of two types:
  Those who know that they've been hacked and those who *don't*!
- Take the *best level* of precautions that you can. Act as if there were an epidemic due to a dangerous disease loose in the city. Employ tools, processes, and training in every area and at every level—from devices like laptops and mobile phones to backend servers and websites.
- There needs to be both a mitigation plan (for prevention) and a contingency plan (recovery from an attack).
- Prepare a response plan organization-wide, as if a cyber attack has *already* happened, and you need to recover from it. Part of the response should always be to use backup of your critical digital assets to restore the business-as-usual status.
- It is vitally important to classify assets, such that the few critical assets are always identified and backed up regularly.

7.9   MASTER COMPLEXITY

The ICT revolution is largely powered by the internet, mobile telephony, and smart devices. Again, within mobility, it is mobile data which is driving the business models. And the rise in devices is fueled by computing power and storage. Not only are the devices proliferating, but their connectivity to each other and to the backend systems is causing an unprecedented growth in business models in the form of cloud, social media, analytics, and mobility.

How does an organization benefit from all these, for both improving efficiency and producing new innovations?

From the social capital point of view, first there are several salient points to consider in this revolution:

1. Understand the difference between complex and complicated. Complexity refers to intricately interconnected subsystems, often having overlapping concepts and patterns, which make it dense. It is easy to understand a complex system if one understands the pattern of interconnections, the subsystems, and how they inter-relate to one another. There are rules and philosophies that guide the organization and development of a complex system. The human body is a wonderful example of a highly complex system. A complicated system, on the other hand, implies a mess. There are no rules or patterns that guide the understanding or development of such a system. A slum is an example. Complicated systems must be avoided at all costs, as they are expensive to operate, impossible to understand, manage, and scale, whereas complex systems are immensely useful because they allow one to achieve more with minimal effort, if managed properly. A system grows from simple towards complex for a reason. If not managed well, it is easy for a complexity to slip into complicated mess.

2. Complexity has, in general, increased enormously. There are layers of components coupled with each other. The versions of the different components need to integrate with each other. There will be ripple effects across the ecosystem of the organization due to changes. Hence the entire organization needs to be prepared to handle changes brought about by the complexity.

3. The speed of change has increased, further complicating the situation. There is little reaction time for people to get trained and absorb the changes.

4. The hidden complexities cause defects and problems which cause serious business risks, and crises that the organization is ill-equipped to handle.

5. There is the question of living with partly new and partly old (legacy) systems, which again complicate the processes and also increase the risks of failure and security hazards.

6. On the other hand, the management and other stakeholders' expectations are huge. They expect the organization to leverage the new changes and profit from them, because a lot of investment goes into the new technology and transformation. There are also constant comparisons to the competitors and other industries that have used the new technologies. Above all, they expect new levels of efficiency and even innovations from the new technologies.

There are several ways to deal with and benefit from the complexity:

1. *To Simplify, Embrace Asymmetry*: Complex systems often follow the Pareto principle (80/20 Rule). Hence, focus on the vital few drivers that make a huge difference. Realize that with technology, the equations can be much more skewed. For example, introducing a new technology, such as mobile applications for your workforce in the field, can drastically tilt the game in your favor. This mindset change is a conscious and constant process, and it should lead to the development of measurement systems and analytics which reflect the principle of asymmetry.

2. *Be Customer-Centric*: Align the entire organization to be customer focused. Measure everything against customer needs. With there being continuous changes in the industry, business, technology, competition, and so on, the one constant for a business is the customer. The customer needs may also keep changing, but the organization cannot exist without customers and stakeholders—hence the need to be keenly aligned to the customer. There should be clear-cut, updated customer dashboards that measure and showcase the critical parameters that are of priority to the customer. It is best to automate and offer self-service portals to the customer, to provide superior and flexible service round the clock. A product management function that helps craft superior customer experiences can help drive this across the organization.

3. *Study* the competition, the industry, the regulations, and similar industries/organizations: Learn constantly, draw parallels, and make your moves, innovating as needed. One can learn a lot by merely reverse engineering, copying, tweaking, re-modeling, and re-engineering. As emphasis on services grows, the strength of product IP weakens; hence, it is easier to copy, without legal impli-

cations. However, note that this applies in the other direction too. Remember, the competition is studying you as well.

4. *Think Like a Startup*: While there need to be special technology/ R&D/digital media teams/labs, the new order dictates that every team must be prepared to experiment constantly and rapidly learn from mistakes and failures. Favor lean/agile frameworks and processes. Experimenting, prototyping, and figuring out what works best is the surest path to growth.

5. *Set Up Hackathons*: These are structured bursts of collaborative innovation that cut across teams and/or organizations and bring people together in one place (physically or virtually), to crack a defined problem in many different and unique ways. These typically last for a day or several days.

   Examples: Topcoder (www.topcoder.com), Skillenza (https:// skillenza.com)

6. *Set Up Digital Platforms and Build Ecosystems*: Expose interfaces, such as APIs, such that different teams can experiment, prototype, contribute, and learn in their own unique ways. Provide tools, access, and spaces to help build ecosystems.

7. *Standardize components/processes and Integrate*: Once proven to be working, the components and the attendant processes should be standardized and integrated into the main ecosystem. Constantly re-factor to simplify and eliminate waste. Avoid fragmentation.

8. *Organize*: Layer and segment the architecture. Apply systems/ design thinking from the beginning to get it right. Get the right fit for the architecture.

9. *Automate, Automate, and Automate*: There should be a constant drive to reduce and eliminate manual processes. In fact, automation should be on the agenda right from the start. While any idea manifests itself, the foregone conclusion should be that this innovation when implemented will be entirely automated.

10. *Set Up Conventional as well as Unconventional Training and Learning*: Training should be done through both normal methods, such as classrooms, labs, workshops, one-on-one teaching, online courses, and so on, and through offbeat methods such as gaming, social media, AI bots that help teach and correct mistakes, virtual reality, augmented reality, simulation, and so on. Tie up with industry and academic forums to get help with training. For instance, NASSCOM in India has launched a future skills initiative

that will train employees and students in any of the identified set of skills associated with an identified set of future job roles that the industry will need. This is described in another chapter.

11. *Use Open Innovation*: Always be ready to collaborate, connect, and co-create (Bigliardi and Galati 2012), especially when the co-creation is done with a player who has a niche that is far removed from your main business, and the collaboration has high potential value.

The Takeaway Box: Complexity Is a Good Slave but a Bad Master! Adopt "Simplexity"!

- "Simplexity" is a mindset that has simplicity as its primary goal and will allow any measure of additional complexity if and only if it helps greatly simplify the overall picture. Hence, simplexity makes complex things simple so that employees and stakeholders can easily understand it.
- Master and leverage complexity or it will destroy you. Entropy is too powerful. Things tend to unravel fast.
- Employ experts in the right places at the right time to help you identify the important factors that you should be focusing on.
- Do not commit investments on a large scale until all key aspects are proven on pilots and prototypes. Encourage hacking to crack problems in new ways. Use innovative training methods to ensure that concepts and practices sink in.
- Study different ways to not only deal with complexity but benefit from it.
- Systems and processes must help you simplify how you deal with and leverage complexity, not complicate things further.
- It takes coordinated energy, effort, and intelligence to keep a complex system together over time; it needs to be built into the culture.

7.10   ACHIEVING SYNERGY

Synergy is the epitome of best practice in an organization. It is the source of great many innovations and positively indicates a high social capital in an organization. Unfortunately, it is most neglected and often underestimated. By definition, synergy means that the sum is greater than the parts.

This means that even a simple idea when combined with another simple idea can give rise to a brilliant idea whose power is a millionfold greater that the other two ideas taken individually. The question is how does one start such a chain reaction? Again the answer is simple, but yields magic when really put into practice.

Synergies are discovered and realized only when people find common interests and work together beyond established boundaries. It cannot be done when they are mentally stuck in silos of any kind. Hence, it is essential for the organization to provide the means for people to interact and discover common problems to solve and find mutually beneficial solutions.

The first step is to help people discover information, data, and knowledge both within and beyond their silos. Searchable knowledge bases, repositories, and tools greatly help. Product management drives the features of a product through requirements engineering and helps identify areas that the customer cares about. Mapping customer journeys, storytelling, interviews, and focus groups help bring out the key features that the teams can focus on. This entire exercise brings about a cohesiveness in the teams.

The second step is to provide forums and incentives to bring different types of people together. This can be through training programs, discussion forums, online forums, forming special interest groups that cut across boundaries, rewarding innovations, hackathons, problem-solving contests, and so on. Sometimes even peer reviews and audits by people from other parts of the organization help either party in understanding how other groups think and function. This results in cross-pollination of ideas and best practices.

The third step is to document and measure the outcomes and pursue the best ones. Good synergistic ideas must absolutely be implemented without delay. The organization must be flexible enough to be able to create new teams in the organization which will implement the new ideas. These can be task forces, or temporary project teams, or permanent ones.

Synergy does not merely refer to ideas that result in products, solutions, or services but also the processes. Innovative organizational practices when standardized can bring about a great synergy between teams, thereby increasing the capability of the organization manifold times. However, one needs to take care that the synergy does not cause the teams to lose flexibility because they're rigidly tied together in the name of efficiency and synergy.

7.11   The Macroeconomic View

At the national level, building social capital needs a different framework. Many countries have now recognized the importance of consciously building their innovation capacity, rather than taking a passive approach of letting innovation grow as a side effect or consequence of the developments in infrastructure, technology, and education. They encourage innovation clusters, such as those described in the chapter on strategy. For example, Food Valley in Netherlands is a large cluster of agro-tech startups and experimental farms. Its nodal point is WUR (Wageningen University & Research), which is considered the world's topmost agricultural research institute (Viviano 2017). There are many ways to score the status or progress of a country in terms of technological and innovation capability. There are indicators for measuring the infrastructure, technology, and education levels of the population of a country. The measures, global competitive index, global innovation index, and network readiness index, all have components that measure the innovation infrastructure. These are given in the chapter on regional factors.

Broadly, the three different components of a national innovation infrastructure are given in the table below.

S. No.	Component of the innovation infrastructure	Description
1.	Technological sophistication	A simple measure is that of number of technological utility patents filed in a country in any year.
2.	Social and financial capital	Resources available for research and development and adaptiveness to technology are the indicators. The quality of connections in the ecosystem is crucial.
3.	Resources committed and policy framework	The budget layout and the policy framework dictate the future of a nation's progress.

Source: Adapted from Roberts (2002)

The national innovation infrastructure shapes cluster specific environments for innovation, for example, technology startup hubs, ICT forums, agro-innovation hubs, co-creation facilities, and so on. This sets the environment as regards context for firm strategy and competition, the input conditions, the demand conditions, and the supporting industries such as vendors, suppliers, and partners (Roberts 2002). Thus, a nation can foster the ecosystems and culture for innovation in a nation. This social capital greatly helps in the progress of the country, regardless of the micro-level changes that happen. For example, if a startup firm shuts down, its employees carry their knowledge, skills, and experience to another firm in the ecosystem, or start another company (with all their valuable lessons learned) in the same ecosystem after some time. So, the ecosystem becomes a growing, learning environment.

The Takeaway Box: Synergies Are the Crown Jewels of Social Capital

- Your main goal should be on how to make a ten-people organization function like a 1000-people strong organization. Increasing synergies and innovating is the answer. Automation is part of that equation and helps speed up things.
- A simple idea borne of synergy and applied in the right place at the right time is far more powerful than a thousand ideas neglected, such as unused patents lying around.
- Synergy comes about only when different thought streams come together and people of different backgrounds work in common interest towards a goal.
- It is very difficult to realize any synergy when systems and organizations are fragmented.
- Integration and shared vision are key for achieving synergy.
- IP (intellectual property) is increasingly about the speed and quality of innovation. It is not a static treasure that you lock and keep. It is very dynamic, organic, and changing. This implies that, while legal patents and IP will continue to have value, in many cases the real IP of new age lies in the social capital of an organization.
- Innovation, especially open innovation, needs experimental, open, innovative minds. It is about creating future business. Hence, experiments, risks, and failures should be encouraged as a perfectly natural process. It is not about generating today's sales. There is no point in getting great sales for today by destroying tomorrow's business. The leadership needs to recognize this.

7.12   Conclusions

Customers' needs are diverse. They differ in terms of their habits, geography, socioeconomic backgrounds, goals, risk appetites, and circumstances. It takes a cross-disciplinary view of domain, technology, business, and customers to relentlessly create value. This is possible only through a complete shift in mindset and culture of the firm towards ambidextrous thinking, intrapreneurial attitude, and a holistic view of the organization. This social capital needed to be able to create diverse and trusted partnerships and collaborations. It also needs the ability to take risk with different strategic initiatives, involving time, money, resources, and collaborations. The organization should have the capability to deploy and redeploy resources rapidly; hence, an intimate knowledge of its resource capability and utilization is important. It needs to find different ways to master complexity. The teams should be able to create a variety of combinations of internal and external services to provide a more superior, innovative and complete set of services to the customers. Capturing value in terms of both intellectual property and collaborations will help a firm secure its future. Synergy is a worthwhile goal that every organization should aim for.

Social capital components capture benefits and activities associated with social networks. This is linked to the softer elements of innovation mostly in the form of social benefits of innovation and knowledge absorption, transformation, and creation. Studies found that strategic "network capital" facilitates innovation. Socialized interaction help in the spillover of economic benefits. However, critics feel that too much focus on social capital may reduce the resource flow which is required for innovation (Murphy et al. 2016).

References

Bigliardi, B., Dormio, A. I., & Galati, F. (2012). The Adoption of Open Innovation within the Telecommunication Industry. *European Journal of Innovation Management, 15*(1), 27–54.

Birudavolu, S. (2015). Open Innovation in ICT: An Empirical Assessment of Global Telecommunications Services. Unpublished thesis submitted for PhD at Indian Institute of Foreign Trade, New Delhi.

Fasnacht, D. (2009). *Open Innovation in the Financial Services: Growing Through Openness, Flexibility, and Customer Integration*. Berlin: Springer Verlag. eBook ISBN 978-3-540-88231-2, Hardcover ISBN 978-3-540-88230-5, Softcover ISBN 978-3-642-09996-0.

Hamel, G., & Prahalad, C. K. (1996). *Competing for the Future*. Harvard Business School Press, ISBN-10: 0875847161, ISBN-13: 978-0875847160.

Looper, Christian De. (2017, December 24). Apple vs. Qualcomm: Everything You Need to Know, Digital Trends. Retrieved from www.digitaltrends.com/business/apple-vs-qualcomm-news.

Martina, Michael and Nellis, Stephen, July 25 2018, Qualcomm ends $44 billion NXP bid after failing to win China approval, Reuters, https://www.reuters.com/article/us-nxp-semicondtrs-m-a-qualcomm/qualcomm-ends-44-billion-nxp-bid-afterfailing-to-win-china-approval-idUSKBN1KF193.

Miller, J. A. (2016, January). The Real Cost Of Unused Software Will Shock You. *CIO*. Retrieved from https://www.cio.com/article/3024420/software/the-real-cost-of-unused-software-will-shock-you.html.

Murphy, L., Huggins, R., & Thompson, P. (2016). Social Capital and Innovation: A Comparative Analysis of Regional Policies. *Environment and Planning C: Politics and Space, 34*(6), 1025–1057.

Roberts, E. B. (2002, April). *Innovation: Driving Product, Process, and Market Change* (1st ed.). Jossey-Bass, ISBN-10: 0787962139, ISBN-13: 978-0787962135.

Sony Pictures Hack. (2014). Retrieved from https://en.wikipedia.org/wiki/Sony_Pictures_hack.

Sun, L. (2017, November). Will 2018 Be Qualcomm Inc's Best Year Yet?, *Motley Fool*. Retrieved from www.fool.com/investing/2017/11/28/will-2018-be-qualcomm-incs-best-year-yet.aspx.

Viviano, F. (2017, September). This Tiny Country Feeds the World. *National Geographic*. Retrieved from https://www.nationalgeographic.com/magazine/2017/09/holland-agriculture-sustainable-farming/.

CHAPTER 8

Regional Factors Influencing Innovation

All differences in the world are of degree, and not of kind, because
oneness is the secret of everything.
—*Swami Vivekananda*

8.1   INTRODUCTION

ICT-based strategies for product/service rollout depend upon regional factors, which include economic, geographical, and cultural differences. As a result, business strategies based on ICT must vary from country to country. A strategy that works in Germany may not work in Brazil and vice versa. A preponderance of regional factors sways both conception and implementation of the strategy. In the previous decades, there was a phase lag between the introduction/use of a new technology in an industrialized country and its eventual adoption in a less economically developed country. With the widespread use of the Internet, most technologies are available or visible to everyone today. However regional factors such as infrastructure, regulations, education, political stability, affordability play into the mix, and affect the outcomes. It is also important to note that just like production value chain, there are now international value chains for innovation, where some parts of the innovation are outsourced. In some cases, non-critical parts of the innovation are outsourced from the emerging developing world where the cost of innovation is quite low. Also, due to geographical, cultural, and regulatory differences, there are strong localization needs for products in some of the large developing countries

© The Author(s) 2019
S. Birudavolu, B. Nag, *Business Innovation and ICT Strategies,*
https://doi.org/10.1007/978-981-13-1675-3_8

such as Brazil, China, India, and so on. Local industries are very active in this localization process which gather domain knowledge and help OEMs to conduct a bigger innovation. Growing regionalism in different parts of the world pushed for such collaborations.

As mentioned above, a strong demand for localization forces the products and services to be offered differently in different countries. As the long tail phenomenon becomes all pervasive (Anderson 2009), the rich local cultures and their unique attributes thereof gain strength and expression through affordable technology. ICT can happily breathe life into dying arts, minority cuisines, traditional attires, and so on, which are languishing and shrinking due to economic recession and lack of market access. Localization, online retail, B2B e-commerce, aggregators, location-based services, and even plain blogs are just a few means to empower the regional ethos.

Technologies themselves benefit greatly from deployment in different countries. A feature-rich enterprise software developed for telecom in Canada could be very successful in that region. Hence, it could be considered for use all over the globe. But when deployed in India, the software may collapse under the strain of tens of millions of users hitting the system every hour. The load is on a scale that is unthinkable in Canada or USA; hence, the software wasn't designed for this level of stress. The software will need to be improved and upgraded for performance, scalability, and unique features, based on the detailed technical feedback from the Indian deployment. This will benefit its prospects globally and raise the bar for similar software in the industry. Hence, the Indian experiment seems to be a useful feed in the innovation process, and perhaps a few Indian companies can also cooperate with Canadian companies in this regard.

A few other examples from India are: a variety of complex language fonts within one country, stringent regulations of SEBI (Securities and Exchange Board of India), and BOP (bottom-of-pyramid) business models.

The biggest boon arising from the regional implementation of technologies is the bounty of innovation that blossoms from the sheer variety bestowed by the different circumstances, geography, history, language, tradition, paradigms, mindsets, cultures, and affordability across the world. People residing in a city that is on the seacoast have a different lifestyle than those in a city in the desert regions. The weather patterns result in stark differences in their living, business, and leisure, from their cuisine to the risks and precautions they take and their general habits. All these result in interestingly different problems and solutions thereof. For example, a problem like frequent food spoilage due to humid weather may not be a chief concern in a drier climate.

Exploiting regional factors for innovation and business success is the subject of this chapter.

Box 1 An Example of Collaborative Innovation: ICT and Development Issues

Intel launched a program called Intel Involved Social Initiative Contest in August 2009. In partnership with an Indian NGO called Mythri Sarva Seva Samithi, it helped volunteers build a platform called Swachh Map, which is a crowd-sourced tracking tool that aims at mapping all the garbage dumps and dirty sites across the country and track them with color coding on a Google Map (Swachh Map 2018). The chief minister of a state and the top government officials of the state and the city can quickly look at the specific localities that need urgent attention, and deploy resources, track progress, and take relevant action. So far Swachh Map has enabled 80+ cities, has 145,000 downloads of its app, and has 27,000 spots cleaned up. It has the full support from governments—owing to the Swachh Bharat initiative—and from the citizens who benefit immensely. (Source: swachhmap.com)

8.2   REGIONALISM AND INNOVATION: ECONOMIC AND STRATEGIC DIMENSION

Japan's effort to fragment the production process and outsource parts and components is noteworthy. As an example, we can talk about the development of Thailand as an offshore export platform for Japanese companies in the automobile and electronics sectors. The cost of production in Japan is prohibitively high. At some point of time, Japan was fast losing its comparative advantage in automobiles and electronics due to its rising production cost. Southeast Asian countries such as Thailand were chosen by a few Japanese OEMs for developing a production center. Thailand also relaxed FDI rules and introduced business-friendly regulations simultaneously. Japanese OEMs were followed by their component suppliers. They also came with FDI and invested around the factories in Thailand. Over the period, local players started supplying components at the tier 2 and tier 3 levels. Slowly, a technology spillover was observed, and it became defused. Several suppliers from Europe and USA also came to take advantage of the growing demand by Japanese majors. This phenomenon is popularly known as the "flying geese" model in which Japanese OEMs are the "lead goose" and they are followed by other players from home countries. Over time, the cost of production in host countries also starts increasing and then OEMs move to greener pastures to take advantage of the lower cost of production. Technology also becomes more standardized and capital-intensive techniques become more labor intensive. Figure 8.1 explains this in nutshell. For example, the technology for producing a normal color TV is now standardized, and companies producing TVs moved from one developed country to another and finally reaches developing countries for production. However, production processes of HD TVs or hard disks still require critical technology and more capital-intensive production systems; hence, the OEMs are more crowded in Japan or newly industrialized economies (NIEs).

The accelerated dispersion of scientific and technological knowledge and capabilities across geographic borders is happening due to the rapid expansion of international trade and corporate networks of production

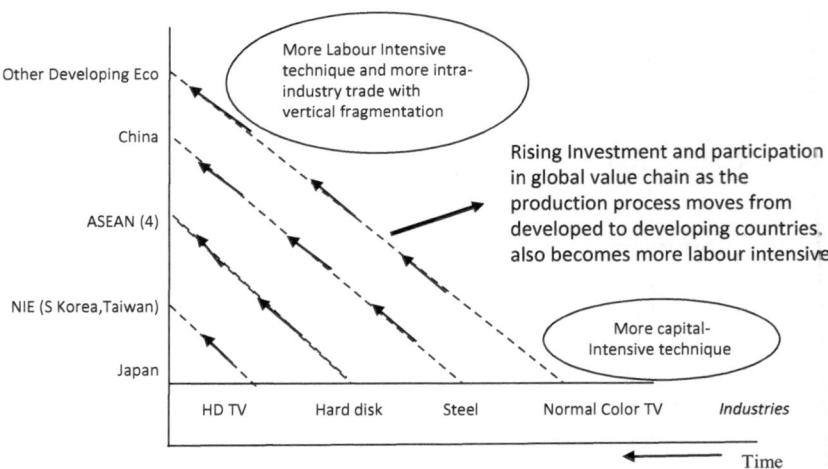

Fig. 8.1 The "flying geese" pattern in Asian economies (Adapted from Kojima, K. (2000): The "flying geese" model of Asian economic development: origin, theoretical extensions, and regional policy implications; published in *Journal of Asian Economics*, Volume 11, Issue 4, Autumn 2000, pp. 375–401)

and R&D. The process is further accentuated due to a liberal FDI regime, improved capacity in absorbing skill and knowledge, and the culture of open innovation. These networks integrate dispersed production, engineering, product development, and research across geographic borders to enhance competitiveness and deep market access. Several trade agreements such as ASEAN and ASEAN+ actively promote a regional production network which also indirectly promotes the proliferation of R&D activities across the region. This has improved regional productivity growth (Brunner 2016). Regional institutions and national innovation policies are the keystones for regional innovation network. Several companies take advantage of this and complement it with the skill and capability of developing countries to spread the innovation activities. Japanese companies are initially backed by national governments for innovation, but eventually, internationalization of the production process drives the success (Nag 2014). In 2007, Phillip Kotler gave a lecture in Jakarta and linked trade agreements with corporate-level international strategy. He gave advices to firms to explore new markets, engage domestic firms, and execute process upgrading as a value enhancer in taking advantage of different forms of trade agreements.[1] Kimura[2] argues that not all kinds of trade agreements promote regional production network. Network development and unbundling of production and even innovation activities happen when network setup costs, service link costs, and production costs go down. OECD has now come out with a database on trade in value added (TiVA). This database can help us understand sector-wise "foreign value added in gross exports". Nag (2016) using TiVA database shows that India's engagement in global value chain has increased substantially in the last two decades, especially in sectors like transport, machinery, and electrical and optical equipment. In 2011, more than 30% of value addition in India's exports from these three sectors are rooted in foreign countries. These value additions mostly come from sectors like trading, financial intermediation, computer and software, and R&D apart from parts and components (outsourcing of manufacturing activities). Globally, along with traditional outsourcing, there is a rise of outsourcing of design, technology, critical components, and so on.

---

[1] Phillip Kotler's lecture on Rethinking ASEAN: Towards ASEAN Community 2015 at the ASEAN Secretariat, Jakarta, August 7, 2007.

[2] Prof. Kimura's lecture at 4th PRI-ICRIER Workshop, Tokyo, March 27, 2014 https://www.mof.go.jp/pri/international_exchange/kouryu/fy2014/kou138_g.pdf.

At the firm level, the strategic choice is to reduce the innovation cost and increase the benefit. Internationalization of innovation activities may get a boost due to institution-driven regionalism. However, the firm has to do a cost-benefit analysis among regional factors that can drive innovation versus the cost due to differences among countries. Standard literature on globalization describes that there are four possible strategies for a firm based on technological and market information requirement. The innovation strategy through an international process depends on these two aspects significantly. This is described in Fig. 8.2. If the need for local market information is high—which means understanding of local market is more important in terms of taste preferences and so on than the technological knowledge—firms need to invest significantly in localization. This is known as a "multidomestic" strategy and in such cases firms have independent units in each country which do product-based innovation to suit the local needs. Many FMCG companies follow such practices. On the contrary, if the requirement of technical knowledge is higher than the need for market information ("global" strategy), firms need much concerted efforts in innovation and it can be carried out in countries where research environment is most suitable. In such cases, firms may think of global products as the need for localization is much less. Apple products

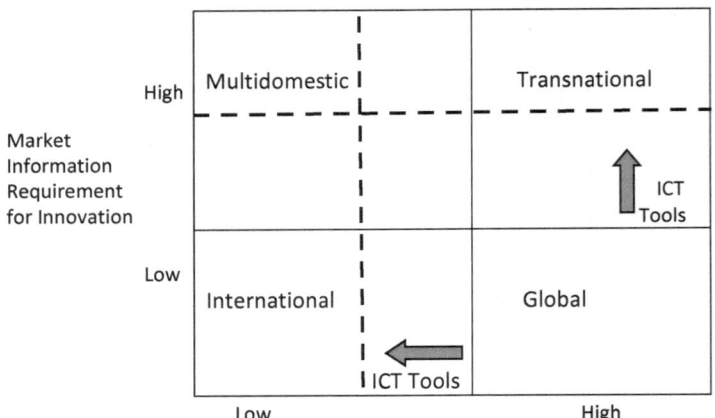

Fig. 8.2 Innovation and global strategy: role of ICT (Source: Adapted from Afuah 2003)

(iPhone, iPad, etc.) could be a good example for this. If both market and technological information requirements are less, firms can use an "international" strategy (Southwest box in Fig. 8.2). Firms can take advantage of local capabilities in different countries without much difficulty as technological innovation requirement is not much. Many international processed food companies use this strategy to localize their products. "Transnational" strategy is required when both market and technological information requirements are high. In this case, firms need to avail the best source for innovation and adaptability strategy—a transnational strategy is costlier and highly risky. The cost of a multidomestic strategy is also not low, and in such cases, companies need to innovate continuously based on regional factors as it moves from one country to another. ICT tools and strategy can help in a big way to manage such situations. It can reduce the need to be physically present in a country to satisfy the local needs (taste and preferences). The new 3D technology can help a company design the products in one place suitable for different countries. Augmented reality can provide realistic situations to design the product as per the local need. It provides an opportunity for some changes in a global product to suit the local requirement and thus reduces the scope for a multidomestic strategy to a great extent. Following Afuah (2003), this can be explained by the shift of the horizontal division of boxes upward (shifting to dotted line from down to up; Fig. 8.2). ICT tools can reduce the scope as local needs and diversified demands can be satisfied without having local production or innovation strategy and hence increase the scope of the "global" strategy (increasing the size of the Southeast box). However, more use of ICT increases the challenge as companies try to be abreast with the latest technology all the time. As ICT is changing very fast and competitors are adopting similar strategies, companies need to continuously invest on technological advancement using ICT. This pushes the vertical division of boxes to the left (shifting to dotted line from right to left; Fig. 8.2). This reduces the scope of a multidomestic strategy further but increases the area of transnational strategy. Hence, a company aspires to be present in different countries. ICT can push the firm more towards a "global" or "transnational" strategy. So, ICT plays a crucial role in managing the regional differences, but it comes with a cost and uncertainty.

The next major question an innovating firm faces is to identify the right place or places for innovation. There may be a host of choices in front of the company, and it requires to choose location(s) carefully. Several indices

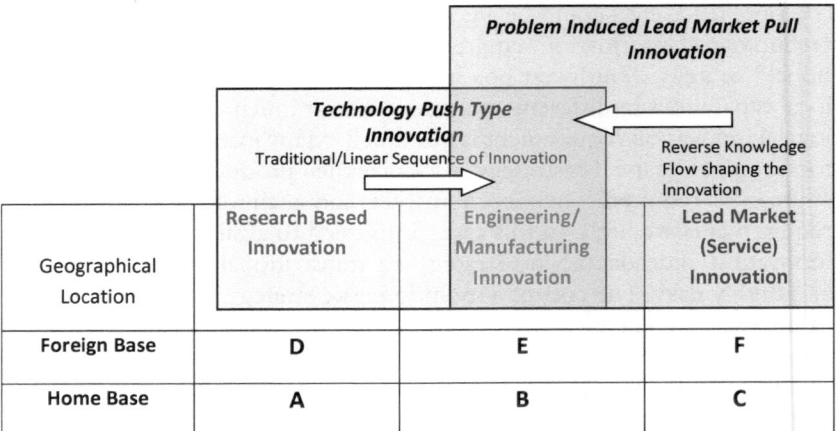

Fig. 8.3 Downstream movement of innovation process, internationalization, and open innovation (Source: Modified and Adapted from Gerybadze 2006)

or economic parameters such as global innovation index, network readiness index, state of competition, and some development indicators give a preliminary idea of identifying offshore location of research activities. This should be evaluated along with economic objectives such as location of and size of the market. Secondly, the firm should be clear about core and non-core value addition in the innovation process and which processes should be procured internationally. In this context, Gerybadze (2006) argues that as innovation is moving downstream, a firm needs to evaluate the location of its R&D with respect to its ability to align with an open innovation model.

Gerybadze (2006) studied the innovation strategy of firms from various countries with respect to their internationalization process. He articulated the framework as given in Fig. 8.3. Traditionally, R&D and product innovation used to be controlled by home base. There has been a linear sequence. First, there is a research-based innovation and then it is followed by engineering innovation to give the final shape of the product using production efficiency and cost optimization. Hence, Home base controls the R&D (Box A) and their manufacturing can be done at home or foreign base (B or E). However, due to rapid growth in competition and focus on market-based innovation, traditional innovation strategy is fast getting replaced by market pull innovation where knowledge of the market pushes the firm to redesign the product or even make a completely new product. There are even examples of later adoption of a product at home base after it is experimented with a foreign base. Hence, there is a reverse knowledge flow. A collaborative effort with local players through joint innovation programs may prove useful in such cases. Gerybadze opined that European, especially German, companies were mostly focusing on home-based engineering innovation (Box B). Japan on the other hand focuses on market innovation, but it has been mostly carried out at home base (C). Both countries have loose connections with foreign bases in this regard. The recommendation was made to both countries to focus on (F) which implies investing on the lead market or service innovation through foreign players. On the contrary, it was found that the USA has strong research capabilities (A) along with strong lead market capabilities but exploits more on home base capabilities (C). However, Scandinavian companies have much stronger capabilities to leverage foreign sources of knowledge along with joint research with local players. Managing innovation with country differences is a challenge, and companies are aggressively using ICT tools and open innovation to address this. The knowledge of national differences and the structural and cultural reasons are the critical factors. The following sections underline some of the strategies which can be useful in this regard.

8.3   THE INDICES FOR REGIONAL FACTORS

Regional factors should be considered in crafting or improving any strategy. Oftentimes, during innovation or in playing out a response to a competitive strategy, an organization may borrow a successful idea from another geographic region. The idea is indigenized for local deployment. Or that "foreign" idea may be combined with another that is a local idea or even an existing local practice. In any case, it's good to study the origins of the idea to gain a better understanding of how to use it. This can lead to better innovation. Borrowing and implementing ideas from similar ranking regions usually poses lesser challenges than from those regions that have a widely different rank.

But what are the regional factors? How do you get a handle on the differences between regions in an objective manner without getting lost in a complex interconnected maze of topics? Each topic is a subject of unfathomable depth, for example, economics, culture, geography, history, politics, and so on. There is a risk of stereotyping and bias if strategies are drawn up based on the experiences and mindsets of a handful of people. Thus, it is imperative to gain an objective and comprehensive understanding while working closely with experienced people and local people.

The following list of indices developed by the industry and government bodies around the world help highlight regional differences. Each of the indices is constructed from a set of variables, which are themselves worth studying. Depending on the specific area of interest or business problem, it is good to research the right indices and their values pertaining to the geographic regions being researched. Different businesses and organizations will obviously be interested in different indices/metrics.

In general, it is good to know the lay of the land of where the innovation or technology is being targeted for *deployment* as against where it was *developed*. Many failures can be attributed to a lack of understanding of the regional market (Table 8.1).

Depending on the business in question, one should pick the right indices to craft a unique strategy based on the differences. This should be part of the overall market study and research. The innovation strategy, especially, is quite sensitive to regional variations. In general, both the strategy and the execution should have a clear regional component. For example, if the inclusive development index and the Gini coefficient of a country indicate that there is very high income inequality, one may consider launching two variants of a service, for example, basic and premium, with high price difference.

A startup generally does not get into a top-down analysis, as it has limited resources and time. It is more focused on a specific problem that it solves in the industry, and builds its business value proposition around the solution it offers. But once it has cracked the basic problem and gets a few customers, it will need to scale up. And this is when it can start exploring the larger picture. For instance, it may discover that it can find market access into certain geographies where its solution will have appeal.

Table 8.1 Economic, technology, and livelihood indices to understand regional differences

#	Name of the index	Developed by	Description
1	GII—Global Innovation Index	WEF (World Economic Forum)	Ranking of Countries by their capacity for—and success in—innovation, which is one of the 12 pillars of competitiveness. It is based on 82 indicators, covering political environment, education, infrastructure, and business sophistication. Published annually.
2	GCI—Global Competitiveness Index	WEF (World Economic Forum)	It ranks countries overall by combining 113 indicators grouped under 12 pillars of competitiveness: institutions; infrastructure; macroeconomic environment; health and primary education; higher education and training; goods market efficiency; labor market efficiency; financial market development; technological readiness; market size; business sophistication; and innovation. Published annually.
3	NRI—Network Readiness Index	WEF (World Economic Forum)	A key component of the GITR (Global Information Technology Report). The report assesses the factors, policies, and institutions that enable a country to fully leverage ICT for increased prosperity and crystallizes them into a global ranking of networked readiness at the country level in the form of the NRI. Published annually.
4	ICT Development Index	United Nation's ITU (International Telecommunication Union)	The Index is based on the internationally agreed-upon 11 ICT indicators, grouped in three clusters: access, use, and skills. Published annually.
5	Global Talent Competitiveness Index	INSEAD and Adecco Group	This Index measures countries' ability to attract talent. Connectedness is an important attribute in GTCI as it effects transformational change by building a strong ecosystem energized by active public-private alliances.
6	World Investment Report (Annual)	UNCTAD (United Nations Conference on Trade and Development)	The World Investment Report focuses on trends in foreign direct investment (FDI) worldwide, for 196 economies, at the regional and country levels and emerging measures to improve its contribution to development.
7	Herfindahl-Hirschman Index (HHI)	Calculated per sector and per country	HHI is a commonly used measure of market concentration. A high value indicates monopoly and a low value, fragmentation or high competition.
8	UN Habitat Urban Data	UN Habitat Reports	The data/coefficients/indices for cities made publicly available by the UN Habitat. (UN Urban Habitat Reports)

(continued)

Table 8.1 (continued)

#	Name of the index	Developed by	Description
9	Gini Coefficient	Widely used, for example, in OECD Reports	It is used to measure inequality. Technically, it's a measure of statistical dispersion intended to represent the income or wealth distribution of a nation's residents.
10	Diversity Indices	Used to measure different types of diversity in a region. By different organizations	UNESCO World Report on Cultural Diversity, Esri Diversity Index, Simpson's Diversity Index, Shannon-Wiener index, Fearon's Linguistic Diversity, Alesina's Fractionalization Scores, and so on.
11	Safe City Index	Economist Intelligence Unit	Cities are global social and economic hubs, and their safety is paramount for the progress of a region. It ranks 60 global cities across 49 indicators covering digital security, health security, infrastructure security, and personal security.
12	Global Liveability Ranking	Economist Intelligence Unit	This rating quantifies the challenges that might be presented to an individual's lifestyle in 140 cities worldwide. Each city is assigned a score for over 30 qualitative and quantitative factors across five broad categories of stability, healthcare, culture and environment, education, and infrastructure.
13	Global Democracy Index	Economist Intelligence Unit	The index is based on 60 indicators grouped in five different categories measuring pluralism, civil liberties, and political culture. In addition to a numeric score and a ranking, the index categorizes countries as one of four regime types: full democracies, flawed democracies, hybrid regimes, and authoritarian regimes.
14	Global Hunger Index	IFPRI (International Food Policy Research Institute)	The GHI is a multidimensional statistical tool used to describe the state of countries' hunger situation on a 100-point scale. GHI combines four component indicators: (1) the proportion of undernourished as a percentage of the population, (2) the proportion of children under the age of five suffering from wasting, (3) the proportion of children under the age of five suffering from stunting, and (4) the mortality rate of children under the age of five.
15	GaWC Index	Globalization and World Cities Research Network, UK	The GaWC examines cities worldwide to shortlist 307 world cities, then ranks these based on their connectivity through four "advanced producer services": accountancy, advertising, banking/finance, and law.
16	Global Militarization Index	BICC (Bonn International Center for Conversion)	The Global Militarization Index (GMI) depicts the relative weight and importance of the military apparatus of one state in relation to its society as a whole.

	Name	Source	Description
17	Military Strength Index	Credit Suisse	The factors under consideration for military strength and their total weights are number of active personnel in the army (5%), tanks (10%), attack helicopters (15%), aircraft (20%), aircraft carriers (25%), and submarines (25%).
18	The Wealth Report	Knight Frank	The Wealth Report picks locations with a potentially bright future, in terms of wealth creation opportunities.
19	Soft Power 30 Index	Portland Communications and USC Center on Public Diplomacy	Soft power describes the ability to attract and co-opt rather than coerce (hard power), using force or giving money as a means of persuasion. Soft power is the ability to shape the preferences of others through appeal and attraction by means of culture, political values, and foreign policies.
20	HDI (Human Development Index)	United Nations Development Programme	HDI is a composite statistic (composite index) of life expectancy, education, and per capita income indicators, which are used to rank countries into four tiers of human development.
21	CPI (Corruption Perceptions Index)	Transparency International	Annual ranking of countries by perceived levels of corruption. The CPI generally defines corruption as "the misuse of public power for private benefit".
22	GNI Nominal (Gross National Income)	World Bank	The Atlas Method from World Bank
23	PPP (Purchasing Power Parity)	IMF (International Monetary Fund)	PPP economics compares different countries' currencies through a market "basket of goods" approach. Two currencies are in equilibrium or at par when a market basket of goods (taking into account the exchange rate) is priced the same in both countries.
24	Global Peace Index (GPI)	Institute for Economics and Peace (IEP)	GPI ranks countries around the world according to their peacefulness. It is a measure of the relative position of nations' and regions' peacefulness. The index gauges global peace using three broad themes: the level of safety and security in society, the extent of domestic and international conflict, and the degree of militarization.
25	Hofstede's Cultural Dimensions for nations	Hofstede's Insights	It elaborates six different dimensions of national cultures: power distance index (PDI), individualism vs. collectivism (IDV), uncertainty avoidance index (UAI), masculinity vs. femininity (MAS), long-term orientation vs. short-term orientation (LTO), and indulgence vs. restraint (IND).
26	Inclusive Development Index (IDI)	World Economic Forum	This index provides a more complete measure of economic development than GDP growth alone. It has three pillars: growth and development, inclusion and intergenerational equity, and sustainability.

The Takeaway Box: Learn How to Use Indices

- Without relying on a solid research base, it is very difficult to get an objective assessment of both regional parameters and where different regions stand on the different scales.
- Research of this nature is nearly impossible for an organization to do on its own.
- Study first the kind of regional differences that your business may encounter, and then research the indices that could be appropriately used. Then delve into the chosen indices to understand how they're constructed.
- Map all these back to the product or service of your business Decide on the relevant changes to be made in the strategy and the implementation
- Conduct special studies and research specific to your business and industry.
- Separate the regional components from the generic product.

8.4   SEPARATING THE REGIONAL COMPONENT IN A PRODUCT AND BUILDING A GLOBAL PRODUCT

Isolating the regional component in a product is always recommended in both strategy and its implementation. This is for an important reason: *building a global product*.

Towards this goal, one should avoid building monolithic solutions that are specific to each country. If you have a solution that was built in a specific region and was very successful, then you need to research well on how to separate out the regional aspects carefully, so that the generic product may be deployed anywhere in the world in conjunction with a local, regional pack ("cartridge") to address regional variations. Internationalization and localization of a product are not merely for the user interface configuration. The effect of different regions is felt in practically every area of the product. This issue has been discussed in detail in an earlier section. The challenge here is how a company is using ICT tools for reduction of technology and market risk and how it reduces the spectrum of multidomestic strategy.

8.5   CHALLENGING AREAS WHEN CONSIDERING A DIFFERENT REGION

Often, there are a few key areas of difficulty that businesses encounter in a new region.

8.5.1   *Availability of Market*

Competition could be very different in a new market. A competitor may have already gained market share and be difficult to dislodge. Sometimes the competition is merely a low-cost and/or low-tech alternative that the population is used to. For example, the biggest competitor to intuit, the software that helps people with doing their taxes, was found to be the ordinary pencil—as in doing taxes on paper forms with a pencil. Training modules need to be built to show the ease of use of the software. Thus, regional habits, preferences, culture, history, geography all play a role. These need to be included in the construction and configuration of the product. For example, in many places in India, vegetable vendors with push carts come right to one's doorstep, and some of the local shops deliver groceries and medicines to the home on a phone call, with cash settlement on delivery. This could pose a competitive challenge to any retail e-commerce company planning to revolutionize retail grocery operations in the style of developed economies. Thus, the entrant companies should adopt a specialized approach for India, focusing on quality or timeliness, or special varieties of groceries, or choosing specific regions that don't have the local facilities. Such a low-level attack would start disrupting the market rather than a head-on conflict.

8.5.2   *Affordability*

This is one of the biggest factors in play. Many markets, like India, are quite price-sensitive. The innovations that work for the masses in developing countries (or similar such sub-regions within the developed countries) are low-cost innovations, bottom-of-the-pyramid, wastage elimination, smaller/affordable packs, economies-of-scale, simplification through standardization, easier/simpler configurations, easy plug-and-play into the larger ecosystem, low maintenance, easy to repair, reusable, low cost spares, extending life through various means, easy upgrade paths, amenable for incremental innovation, working in tougher and often unreliable conditions, fewer models, lesser sophistication, meeting basic needs, and so on.

The innovation themes that may work in the more mature markets or economically forward countries or with the upper (or upwardly mobile) segment in the developing countries are sophisticated, quality conscious, higher end, specialization, fragmentation, online, virtual, connected, flexibility, feature rich, health conscious, lifestyle oriented, comfort oriented, luxury, image conscious, competitive positioning, meeting fine-grained and specific requirements, configuration, personalization, training, learning, and so on.

Examples are easy to come by. Right from vehicles, mobile phones, and household devices to the public facilities and conveniences, the grade and quality of goods/services is comparatively superior in the developed economies. People in the enormously wealthy countries can't even imagine products or services falling below a certain standard. They can also afford high class education and invest in research facilities, which churn out innovation.

Since software will dominate every field, one of the biggest differences will arise due to affordability of high-end software and the training thereof. The case of commercial software is obvious, but even the large open-source software will need customization and training.

However, the developing economies have tremendous advantages that the mature economies do not. In some areas, their expectations may be surprisingly more sophisticated than those in the developed economies. All these are explored in another section in this chapter.

8.5.3   *Expectation of Features*

The so-called developing countries are actually very rich in terms of culture, language, geography, history, society, cuisine, and music and in general a much richer variety in every area. In many developing economies like India, the ethnic wear and jewelry/accessories designs have a much broader range than Western equivalents. The "development" in the Western world has been a very lopsided one, focusing primarily on economy and technology and less on family or culture. The net effect is that over time the advanced technologies will be used more by the developing economies, in terms of quantity, quality, and variety than by the developed economies.

For example, there are 18 official languages in India (with as many scripts) and hundreds of dialects. The television broadcast as well as the internet websites with regional flavors are only poised to explode in numbers and variety. Even though China is one country with one official language, there are many such variations in dialects that they even need translators to understand each other's dialects.

All these indicate that in several areas, the developing countries may have higher expectations of technology much more than the economically developed countries do.

Some more examples:

Mobile banking picked up in Africa is much more than it ever did in the first world countries because traditional banking is weak in several African countries. The developing countries were slow in industrialization hence their energy needs did not grow to the levels of developed economies. Thus, they're not strong in oil lobbies and are badly dependent on others for their energy needs based on oil. Hence, they're aggressively pursuing alternate energies, like solar, electric transport, nuclear energy, and so on. Many tropical countries have abundant sunshine throughout the year; hence, it makes perfect sense. Other examples include: Intricate language fonts that are non-existent in the Western world, learning apps on the mobile for reaching children in remote areas, feature rich phones at lower price points, vernacular language translation, robustness of equipment in the face of heat/dust/wind/rain, fault-tolerant systems, help manuals and tutorials for learning and operating equipment/software especially for self-learning when no experts are around, automation in troubleshooting, and so on.

Developing countries also have the talent to produce A-grade technologies for global needs and export to developed countries. The Indian ICT services sector is a good example.

A specific example:

SenseGiz (www.sensegiz.com) is an Indian product startup company that specializes in IoT technology and has an R&D base in India. Exports constitute over 90% of its business. SenseGiz makes enterprise and industrial IoT products. Its main business is sensor-based condition monitoring, security, and real-time asset/people tracking applications using a combination of proprietary Bluetooth-based mesh connected hardware, cloud, analytics, and applications.

8.5.4   Performance Expectations

The expectations of ICT-driven products and solutions vary from region to region. In India, a million people accessing an online system every hour is quite common. Obviously this is vastly different in other less-populated countries. The performance requirements affect the generic component of the product, often at a design/architecture level. A product from a country with a sparse population would face difficulties in launching its product in a country with high population. But it's best to address these as early as possible, for two reasons:

The first reason is to improve and harden the overall product, and to render it suitable for global deployment, especially in the high-population countries. Almost all products of the top software companies such as Microsoft, Oracle, and IBM have followed this route.

Second, a firm must avoid crafting quick fixes and solutions which do not address the underlying problems. It is necessary to continue to keep the generic and the regional parts separate. Monolithic solutions render future maintenance impossible, and will need an army of people just to keep the solution running. The solution will also resist any generic automation, testing, or upgrades. For example, if issues are found in the generic component at any time and arising from anywhere in the world, then fixes should be made available, such that all the regions should be able to uptake the fixes at the earliest. This is difficult if the generic and regional components are mixed up inseparably. Another advantage is that automation or integration with other products should be picked up smoothly and deployed. The solution should not be bogged down because of lack of separation between generic and regional components, rendering upgrade impossible.

8.5.5   Regulatory Frameworks

Not only are these vastly different from country to country, but they change constantly. On top of it, the governments demand that all companies and organizations show regulatory compliance at all times.

All these requirements should reflect in a product design:

- It is built for regulatory compliance, from the ground up.
- The regional component is cleanly separated, for ease of deployment in different regions, each with its own regulations.

Both of the above are very hard to retrofit and hence should be factored in from the beginning. Some of the common regulations are competition/monopoly monitoring, taxation, environmental, cyber-security, financial, national defense, security and police/anti-crime/anti-terrorism, censorship laws, corporate social responsibility, and so on.

The innovations should not only avoid violating the law, but also *actively* demonstrate compliance in any audit that a regulator may demand at any time. Sometimes, there are fears that MNCs will take away all the benefits, and hence the government imposes barriers or provides preferences to local players so that technology develops indigenously. However, there are pros and cons to this idea. Domestic companies may not have the necessary strength and technological maturity, and they may not be able to cope unless regulation allows good quality collaboration for co-development with MNCs. Hence, regulatory balance is very important.

In the Indian government procurement policy for software, there is a clause for local content requirement (LCR) which tells that 30% of the provided software or hardware should be locally procured. Hence, if an MNC gets the order, they have to procure locally or give subcontract to the local players. However, if the product or service is not available locally, then import can be considered as locally procured.

In some cases, this hurts Indian software companies significantly. As an example, we can see that even after signing trade agreements with Korea and Japan, Indian companies could not get large contracts, especially from those governments.

A chapter in this book explores the regulatory scene in India.

Investments that a government makes in infrastructure and the latest technologies indicate a visionary, progressive leadership. For example, China has committed $150 billion for the establishment of artificial intelligence infrastructure and architecture (Huang 2018), which will help make the country a super power. Regulations and investments go together, guided by a vision.

8.5.6   *Availability of Talent*

This is the single major constraining factor, the importance of which many realize very late in the game. All products, especially new innovations, need considerable talent not only for conceiving and building the product but also for indigenization, marketing, sales, building solutions based on the product, deployment, integration with the ecosystem, maintenance and support, customer training, support staff training, incremental upgrades, project management, integrating, retiring and disposal, continuous improvement, constantly studying the competitors' moves, demonstrating regulatory compliance, managing the supply chain, and so on.

It is easy to underestimate many of these and incorrectly assume that once one has an innovative and brilliant product, all the remaining will follow easily, including revenue and market share. Nothing could be further from the truth.

Many products with great potential have sunk without a trace precisely because their organization failed to build an adequate talent pool, or have tried to push the product in the market in a geography without accounting for the above listed factors. Regularly flying down experts from different parts of the world for every need is not a viable strategy. A solid base of regional talent needs to be built for the success of any product. There need to be many homegrown success stories with a product, rather than grafted or implanted ones. Talents need to be nurtured for the future also. Localization depends on the availability of talents also.

> **The Takeaway Box: Regional Factors Are Key to Building a Successful Global Product**
>
> - Separate the regional components from the base product; the latter should constitute the bulk of the product, to enable easy upgrades, research and development, and maintenance on the core product.
> - The regional components should be like a configurable cartridge, small and distinctly separate. Each region should be able to do its own configuration without overly affecting the base product.
> - Recognize the common difficulties and pitfalls in launching products or services in new regions, and prepare for them upfront.
> - Developing economies are divided into many groups: middle-income countries (Brazil), low middle-income countries (India), and low-income countries (Bangladesh). Only low-income countries are known as LDCs or "third world". They don't offer much opportunity but the other two do. Even so, some ideas can be borrowed from the LDCs also, like the micro loans concept by Muhammad Younus, Nobel Laureate from Bangladesh.

8.6   THE COLOSSAL ADVANTAGES IN DEVELOPING ECONOMIES

Many countries are developing economies which were slow to move into industrialization, and hence their economies remained agriculture based, or had picked up in a few sectors such as mining, tourism, ports, finance, and so on, without going through the full process of industrialization. Examples include several countries in Africa, Latin America, Asia, and the Middle East. As highlighted in the previous sections, innovations and business models in these countries that cater to the majority of the population cannot assume a solid, robust infrastructure on which sophisticated high-end services can be rolled out. Even basic amenities such as drinking water are a challenge in some areas. Saudi Arabia, Bahrain, Brunei, and so on have rich infrastructure and facilities owing to their oil wealth, but are lagging on several accounts—development of alternate economies and industries (other than oil), research and development, creation of institutes of repute, women empowerment, democratic parameters, and so on. Much needs to be done there too, and they've recognized this. The UAE has announced a UAE 2013 AI Strategy and appointed a minister for Artificial Intelligence (Arabian Business 2017). They intend to quietly switch from oil to AI, especially in light of the uncertainty in oil prices. For the majority of the developing countries, there are a few tremendously huge advantages that these countries have over the developed economies:

1. **Ability To Skyrocket, Not Just Leapfrog**
   The developed economies have sunk trillions of dollars into their infrastructure over the past decades. While the benefits are obvious, this is also a double-edged sword. Their infrastructure also locks them down into known and older patterns and models. Upgrading is difficult and expensive, for instance, for all the old COBOL-based mainframe systems that still run Wall Street businesses in New York. However, the newly emerging economies have no such constraint, as legacy systems are practically non-existent and can thus directly adopt the latest, best-of-breed, and low-cost technologies. For example, instead of wiring up an entire building to provide LAN connectivity, they may simply install a few latest Wi-Fi routers at key points to hook up everyone instantly.

The new economies need not traverse the same path that the advanced countries took!

They can strike out an altogether new path to the future based on the new generation technologies that are far more lean, efficient, and economical. The best part is that the new technologies are likely to be standards-based and accompanied by and in fact capable of being integrated well into larger models and entire ecosystems. With the technologies and business models evolving at warp speed, what is required is not just leapfrogging, but skyrocketing. This implies that not only are these developing economies able to leverage the latest technologies as explained above, but they are able to actually create them and combine them in novel ways. Examples of skyrocketing in India include India's research programs in space, nuclear, communications like NAVIC (GPS-like system), missile technology and defense research, India Stack, and so on. It also includes regulatory measures such as SEBI guidelines, enforcing net neutrality, demonetization, and the push towards digitalizing, making generic medicines available, setting up startup incubators and startup growth engines like T-Hub, and so on.

2. **Predominance of Youth**

   Developed economies like the USA, countries in Europe, Japan, Singapore, Australia, New Zealand, and so on have an aging population, owing to a stationary or declining birth rate. Their youth population is effectively diminishing, and many of these nations are now dependent on a regular infusion of capable youth through immigration from other countries to be able to sustain their economy and living. However, many of the developing economies, barring those suffering from severe war or famine, are rich in human resources, are blessed with a large youth population, and have sustainable birth rates. The youth constitute the bulwark and can propel their economies through work, learning, talent, collaboration, innovation, and consumption. The youth work to support the rest of the population. Here, the developing countries have a definite edge over the more mature economies; the latter will soon be dependent on the former for talent and human resources (i.e. youth). Youth are producers, youth are consumers. For example, youth both create and use mobile applications, data, content, and resources far more than the elderly and retired do. The youth are the working population, and they innovate and create the present and future in a far greater measure than the elderly do. The youth are the future of any country.

3. **Redefine Basics**

   With all the constraints that the developing countries have, they're compelled to look at everything very differently than their counterparts from the developed world. As a result, they will redefine the basics. The mature economies have long propagated a society and lifestyle that is basically unsound and unsustainable. The developing countries cannot afford to go down that route. Hence, they will relook at technology, its uses, its challenges, its side effects, its integration, its costs, the training, and so on in a completely new light. This gives rise to fundamentally new models in every area. For example, the Western world was built on proprietary software over the past half a century, whereas the developing countries will largely use open-source software. Other examples are micro-credit banking by Grameen Bank in Bangladesh and elsewhere, adoption of generic drugs in healthcare, move towards alternate therapy/medicine such as Ayurveda, Yoga, Unani, Siddha, Homeopathy, and Naturopathy (AYUSH), and so on.

   An Example:

   Health Cube is innovative and integrated low-cost medical diagnostic kit that can be carried in a handbag (Health Cube 2018) that does 30 medical tests. It offers quick diagnostics with quality results, which is a boon to people in remote, under-developed areas. It needs minimal training and minimal maintenance. The tests are conducted within minutes and it is cheaper than the conventional diagnostic tools. There are wide applications for Health Cube, ranging from social responsibility, insurance, entrepreneurs, corporate, governments, consumers, public transport, and medical establishments. Health Cube is perfect for rural India and is indeed.

4. **Brilliant Innovations**

   In general, new insights and innovations arise from many sources, such as:

   (a) A variety of different needs
   (b) New and different circumstances
   (c) Multiple kinds of constraints
   (d) New combinations, connections, and cross-pollination
   (e) Dividing and dissecting
   (f) Applications of new concepts
   (g) New applications of old concepts
   (h) New collaborations
   (i) Incremental improvements

   Regional influences enable all of the above, as follows (Table 8.2).

Table 8.2 How regional factors power innovation

S. No.	Source of innovation	How the regional factors power innovation
1	Variety	Great variation due to abundance of variety in the different regions: geography, history, culture, society, language, traditions, flora/fauna, economy, politics, development, and so on.
2	Circumstances	Circumstances shape a person's thinking in unique ways, which is a fertile ground for new ideas and innovation. Covers a wide spectrum for people across the world. There are many categories, for example, rich/poor, young/old, urban/rural, healthy/sick, educated/uneducated, intelligent/ordinary, working/unemployed, native/immigrant, from a developed country/from third-world country, from a peace zone/from a war zone, woman/man, business owner/employee/self-employed, employed with government/with private sector, married/unmarried, technical/non-technical, and so on. Thus, there are specialized products for the elderly, infants, and so on.
3	Constraints	Constraints are a way of spurring innovation. Constraints impose reduced options. Necessity is the mother of invention. Constraints can take a million forms, for example, physically challenged, health problems, lack of time, lack of access to education, water scarcity, affected by natural disasters, lack of infrastructure, joblessness, being in a country at war, limited budgets, limited talent pool, severe competition, price war, regulatory requirements, lack of access to information, size/weight restrictions, and so on. Innovation examples include all the wartime technologies, defense tactics that have been used for civilian purposes, studying animals that are hurt regarding how they conserve energy, smaller containers and devices for portability, and so on.
4	Combining, connecting, cross-pollination, synthesis	Combinations and connections can give rise to new value propositions. A few examples are collaboration between international research institutes in different regions, cross departmental research (e.g. nano-robots for medicine), and cross-cultural interactions such as fusion music, fusion food, and so on; when people migrate to another place they combine their knowledge and mindset with the new environment and create new ideas.
5	Dividing, dissecting, analysis	Much insight can be obtained from a complex entity by separating the threads, isolating the elements, and extracting the underlying patterns. For example, natural language processing when applied to languages and dialects from different regions reveals insights about usage, grammar, pronunciation patterns, which when implemented make the system richer and greatly useful. Studying minerals, flora, fauna, geography, and history from different regions helps further science and technology to confirm, validate, and refine concepts and applications.

(continued)

Table 8.2   (continued)

S. No.	Source of innovation	How the regional factors power innovation
6	Application of new concepts	New concepts when applied in different regions can help give better insights from the field, thereby leading to innovation. For example, deploying a new technology product in different regions gets better feedback. A smartphone with new features launched across the globe will get richer feedback. A new medicine can be pitched to different populations across the world, to get data on how it works on people from different regions with different genetics, metabolism, diets, habits, work culture, and so on and concepts of new smart, urban agriculture are discovered in the Netherlands and applied in different parts of the world in order to improve yield while avoiding fertilizers and GM crops.
7	New applications of old concepts	A different region can happily breathe life into an old concept due to varied needs and convenience of use. Examples include age-old therapies (from different regions) for once little known diseases that have suddenly spurted now to become mainstream, electric vehicles suddenly becoming popular/mandatory in different countries because gasoline is non-renewable, polluting, and expensive. Japan took Deming's quality concepts seriously though Deming was an American and was ignored in his home country, USA, and Japanese products conquered the world as a result.
8	Incremental improvements	The bulk of the normal product lifecycle is driven by incremental improvements. Different regions progress at their own pace. Thus, each region is invested in technologies and concepts, right from legacy systems to financial commitments and consumer habits. Rather than replacing everything, they prefer to improve things incrementally and extend the lifecycle at a low cost. Ideas for improvements coming from many regions make the product more robust and enduring. People also learn habits incrementally. This is the essence of the hugely successful Japanese Kaizen approach.

People around the world have been researching, combining, and connecting different concepts and ideas since ages. People at different places combine things in different ways because they have a different worldview, different resources, different needs, and in general a different take on things. Also things occur at different places around the world. Information, expertise, and things are not made available to all the people at the same time. For example, the latest version of technologies and the expertise surrounding them may be available to the Western world, before it comes to the third world, who may find ways to extend the older technologies, find low-cost ways of using and maintaining technology, and putting technologies to new uses. Nevertheless ancient wisdom and sciences, such as Ayurveda, Yoga, and Chinese Medicine have made entry into the Western world only in the past two or three decades.

As is evident from the list above, regional factors and influences are a perfect boon for innovation. With the advent of the ICT, both regional and global influences increase. The regional influences contribute heavily to the long tail, greatly spurring innovation. Some of them can become global trends too, for example, the Indian decimal number system, English words with Sanskrit origins, Japanese Manga style comics coming into the mainstream, a local Thai recipe branded as the Red Bull drink by an Austrian company and sold globally, pizza being made everywhere, and so on.

Evidently, in today's world, ICT plays a key role in making all these happen, in both production and consumption, whether it is for ordering a pizza on a mobile phone, using online tools for creating designs/artwork, or ordering special herbal medicines from an online portal.

An Example:

Artivatic (www.artivatic.ai) is an Indian startup which has developed an AI enterprise tech platform built on genomic science, psychology, and neuroscience capabilities to automate the decision-making, prediction, personalization, and recommendation in real time. It has used its cutting-edge AI platform, tools, and expertise to deliver solutions for banking, robotic manufacturing, and wealth management. For example, it uses neuroscience analogy and genomic pattern-based system in its AI system to take automated decisions on insurance policy cover and approve/reject claims, and even modify the policies on its own, without the need for a human agent. This will be of great help in India, where the population is large (hence a huge number of claims). There are many cases of fraud, and hence, such intelligent automation is a boon.

8.7   Conclusion

The internationalization of the production process is followed by conducting R&D processes in different countries. The increased intensity and speed of innovation, the accelerated product obsolescence, downstream outsourcing of innovation process, more knowledge sharing, and a culture of open innovation have pushed R&D processes to become more global. National firms collaborate with foreign firms for stable market access, localization of product, and further product and process innovation. Regional and national policy for innovation, FDI, patent protection, skill development, and so on play the role of catalysts in this process. A developing country becomes a host of global innovation based on its technology absorption capacity, skilled manpower, innovation culture, large market, infrastructure, regulatory and investment protection framework, and so on. This chapter also explored the diversity caused by geography. It shows why and how the go-to-market strategy of any organization needs to account for it. Indices are a good way to gain a better understanding of the regional factors, especially to compare, contrast, craft, and fine-tune the strategies laid out for different regions. A product/service can be targeted for different regions and should hence have, other than the core product, a regional component also which can be configured/customized for region-specific factors. Some of the key challenges for different regions should be considered, such as affordability, regulations, expectations, and so on. It is best to leverage the advantages and diversity of the different regions to improve the main product itself.

The Takeaway Box: "Developing" Regions Are Highly Developed in Many Ways Than One Can Imagine

- Relative to your product or service, identify the few critical regional factors that will make all the difference for the success of your product in a region.
- The leading pack of the so-called developing economies have the highest economic growth rates, youth population, culture, diversity, vibrancy, and talent that the "developed" countries simply do not possess. In fact, some of these economies were quite prosperous in the past, before they were ruined by invasion, exploitation, war, or natural calamities.
- The future of the world, in every sense, is driven by these developing economies (and the people from these economies) much more than ever before.
- Find ways to leverage the diverse influences from the developing economies. The stories and the thinking from there will spur innovation, productivity, and growth.
- The developed economies must offer investment (capital) and technology to put the world on a track of sustainability and prosperity all around.
- It is futile to pursue the same path that the developed economies took. It is unsustainable.

REFERENCES

Afuah, A. (2003). *Innovation Management: Strategies, Implementation, and Profits.* Oxford: Oxford University Press. ISBN-10: 0195142306, ISBN-13: 978-0195142303.

Anderson, C. (2009). *The Longer Long Tail.* New York, NY: RH Business Books, Random House.

Arabian Business. (2017, October 22). *UAE Appoints Minister for Artificial Intelligence.* Retrieved from http://www.arabianbusiness.com/politics-economics/381648-uae-appoints-first-minister-for-artificial-intelligence.

AYUSH, Ministry of AYUSH, India. Retrieved from http://ayush.gov.in/.

Brunner, Hans Peter. (2016). *Innovation Networks and the New Asian Regionalism: A Knowledge Platform on Economic Productivity.* Edward Elgar Publishing and Asian Development Bank.

Gerybadze, A. (2006). Global Innovation and Knowledge Flows in Japanese and European Corporations. In H. Cornelius et al. (Eds.), *Management of Technology and Innovation in Japan* (pp. 311–327). Berlin, Heidelberg: Springer.

Health Cube. (2018). Retrieved from www.healthcubed.com.

Huang, E. (2018, April). China has become a Technological Powerhouse in Artificial Intelligence. *CNBC.* Retrieved from www.cnbc.com/2018/04/12/china-has-become-a-technological-powerhouse-in-artificial-intelligence-says-global-robo-ceo.html.

Nag, B. (2014). *My World with Rafiki, An Economic Travelogue and Miscellany.* Partridge: Penguin.

Nag, B. (2016). Emerging Production Network between India and ASEAN: An Analysis of Value Added trade in Select Industries. In D. Chakraborty & J. Mukherjee (Eds.), *Trade, Investment and Economic Development in Asia: Empirical and Policy Issues* (pp. 41–67). Abingdon: Routledge.

Swachh Map. (2018). Retrieved from www.swachhmap.com.

UN Urban Habitat Reports. UN Habitat's Online Reports, UN Habitat. (n.d.). Retrieved from http://urbandata.unhabitat.org/compare-cities/.

## CHAPTER 9

The State of the Indian ICT Sector

*If I were asked under what sky the Human Mind has most fully
developed some of its choicest gifts, has most deeply pondered on the
greatest problems of life, and has found solutions, I should point to
INDIA.*
—*Max Mueller*

9.1   Introduction

India is not an emerging economy, but a re-emerging economy. Between
1 AD and 1000 AD, India was the largest economy in the world, controlling between one fourth to one third of the world's wealth. Up to the
seventeenth century, India and China remained the world's largest economies. In 1700, India was the largest economy, producing a quarter of the
world's GDP. India was the most advanced and also the largest economy
in the world for most of the period between 1 AD and 1800 AD. Science,
technology, astronomy, mathematics, medicine, literature, arts, and crafts
flourished in a rich and prosperous India. Many experts consider Sanskrit
language to be the mother of all languages worldwide.

Under the British rule (1757–1947), India was completely destroyed in
every way (economic, social, cultural). India was thoroughly looted
through taxes, exploitation, and through other means such as plundering
raw material from India and importing finished goods into India. The
local trade, science, arts, and crafts were crushed. The net effect was that
Britain became the leading economy of the world and India's share of the

© The Author(s) 2019
S. Birudavolu, B. Nag, *Business Innovation and ICT Strategies*,
https://doi.org/10.1007/978-981-13-1675-3_9

world's industrial output fell steeply from 25% in 1750 to a mere 2% by 1900. The railways, telegraph, and legal system established by the British government were intended for exploitation of India and not for helping the country. As a result, the Indian population was malnourished, and the Indian population had the lowest life expectancies in the world under the British rule. The people were kept illiterate, and famines were frequent, one of the worst being the Bengal famine in 1943 that resulted in ten million deaths (Sen 1981). By comparison, about eight million Jews died in Nazi Germany under Hitler's regime around the same time. India gained independence in 1947, having been split into three parts before the British left. India has since made progress in many dimensions despite severe challenges.

After its independence, India embarked on mixed-economy model, with socialism at its core, and capitalism added in moderation. The landscape was dominated by public sector and a heavily regulated industry, protected from foreign competition by restrictions and tariffs on imports. Oil constituted the bulk of the imports. The balance of payments suffered. Owing to rampant corruption and heavy regulations, poorly implemented and coupled with inept law-enforcement, the private sector suffered. This also resulted in mass unemployment and brain-drain. It could be argued that any successes in the private sector happened despite the government and not because of it (Das 2012). A mounting fiscal deficit reached a crisis point in 1990, when urgent and extreme measures were taken to liberalize the economy. Since then, India has achieved an economic turnaround.

Today, in 2018, India has a GDP of 2.44 trillion USD and is projected to grow at 7.8% in 2018 as per the IMF, making it the fastest growing economy in the world (Panchal 2018). It is the world's fifth largest economy by nominal GDP (Focus Economics 2018), and the third largest in PPP (purchasing power parity) terms (Wikipedia Economy of India). Its GDP has surpassed that of France and the UK.

However, its contribution to the global trade is only 1.7%. It needs to improve its foreign trade exports. For doing this it needs a world-class infrastructure, which is a far cry from its current state. Given that the world is moving towards a digital economy, it is imperative that India embraces ICT.

Unlike most of the economically developed economies, India is also very youth centric. About 50% of India's population is below the age of 25 and more than 65% of the population below the age of 35 (Sharma 2017) (Wikipedia Demographics of India). India is all set to capitalize on its demographic dividend. The youth are the growth engine of India and hence should be trained and empowered for India to move to the center of the world stage.

The consumption class in India has grown rapidly along with the economic growth, leading to greater purchasing power per household. This would translate to a greater portion of income being available for discretionary spending, instead of using up most of the income on bare necessities like food, clothing and shelter.

India has come a long way in Telecom, since the time when telegraph first came to India in 1851 and when it played a key role in the defeat of the 1857 Indian freedom struggle (Vatsa 2012). The British rulers at the time could instantly communicate about the movements of the freedom fighters through the newly deployed electric telegraph network, and thus deploy troops at the right place and time to inflict a defeat on the freedom movement. In 1895, Sir J. C. Bose in Kolkata laid the foundation for wireless communication with his experiments. He also held the first patent worldwide to invent a solid-state diode detector to detect EM waves, and he was a pioneer in the field of microwaves (Aggarwal 2018).

Today India has a full-fledged and successful space program that has put indigenous communication satellites into orbit, thereby carpeting India with low-cost media broadcasting and enabling critical communication for government and defense. The commercial Telecom sector has also boomed rapidly in the past decade.

This chapter will briefly explain the current state of the Indian ICT ecosystem. Along with a brief history of ICT in India, the section will also discuss the various forces that are often at play as far as the ICT ecosystem is concerned: the geographic and socio-economic conditions, cultural factors, business models, the economy, technology adoption, regulations, trends, constraining factors, and so on. From here, specific guidance will be provided to the readers on how to thrive with ICT in the Indian context. For instance, operators taking to direct carrier billing aggressively is one of the ways in which the untapped mobile commerce markets can be seized. The operators can collaborate through open innovation to provide a unified billing interface which they can market to both businesses and the software developers who develop mobile applications. On the regulations side, Telecom and IT have now been split into different ministries of the government in 2016. India's regulatory structure is focusing on network audit, mandatory standards for equipment, more stringent security system, and so on. The relevant section will critically analyze India's alignment with global standards and how a new ecosystem will help business to grow.

9.2   India and its History with ICT

Indian Postal system and Telecom are among the oldest in the world, and functioned under one department until 1985, when they were separated off. Telegraph and later Radio-Telegraph dominated Telecom from 1851 (inception) until the 1930s. Landline telephone entered the scene circa 1914 grew in both usage and development over the decades, for example, a 12-carrier system was introduced in 1953, STD in 1960, and so on. ITI (Indian Telephone Industries) was formed in 1948 to produce Telecom equipment.

Radio broadcasting took off in the 1930s, and All India Radio (later re-branded to Akashvani after two decades) became a very important medium to reach the masses. Television was introduced in the 1950s and grew to complete broadcasting by the 1960s. Both radio and TV were state-owned. The 1970s saw the introduction of PCM (pulse code modulation) in telephones, digital microwave, and fiber optic cables. TCIL (Telecommunications Consultants India Limited) was formed under the DoT to provide Telecom consulting services to developing countries around the world.

In the 1980s, the satellite-earth station was established. Internet was launched with ERNET (Educational Research Network), which was made available to educational and research institutes. On the Telecom side, re-organization resulted in five entities under the state-owned Telecom department, to provide services and equipment to the end-user:

- C-DOT (Center for Development of Telematics): Constituted to develop indigenous R&D and production of digital (electronic) exchanges.
- ITI (Indian Telephone Industries): Telecom equipment manufacturer for switching, transmission, access, and subscriber equipment.
- MTNL (Mahanagar Telephone Nigam Limited) was formed to provide Telecom services to cover the two metros Mumbai and Delhi.
- VSNL (Videsh Sanchar Nigam Limited) was constituted for providing international long-distance telephony.
- DoT (Department of Telecom) catered to the Telecom services needs of the rest of the country.

9.2.1   The 1990s Witnessed the Following Developments

Mobile telephony and the Internet were introduced to the public, and both of them saw explosive growths for the next two decades, continuing to date. The Indian government de-regulated the Telecom sector in real earnest, as part of the economic reforms and trade liberalization policies. These urgent measures were taken to address the surmounting fiscal deficit crisis. The government allowed private players in specific areas, such as mobile telephony and VAS (value-added services). The country was segmented into regions, called circles, which are approximately the same as the geographic regions of the different states of India:

- For basic telephony: 20 circles, with bidding open, in each circle, to one private player apart from DoT.
- For mobile telephony: 18 circles, with bidding open, in each circle, to two private players apart from DoT. Both CDMA and GSM exist, but GSM in the 900 MHz band (and 1800 MHz) is more prevalent than CDMA. Spectrum auctions were held in 1994, 1995, and 1997 to award licenses to operate and use blocks of electromagnetic spectrum in each of the circles in order to provide services.

A 15-year license was given to each selected operator.

9.2.2   Under the NTP (New Telecom Policy), Change Was Mooted in the Three Areas:

- Ownership: Privatization, with a permissible foreign stake of 49%, and multinationals involved in joint ventures and technology transfer, not in policy making.
- Service: Telecom now deemed as a basic service that needs to be made available to all, and hence needed massive expansion into rural India. Joint ventures with multinationals were to be formed for this purpose.
- Regulation: TRAI was formed as an independent body for policy making and determining tariffs. Whereas earlier, the service provider and the regulator were state-owned and a single entity, with possibility of conflict of interest. A separate body TDSAT (Telecom Disputes Settlement and Appellate Tribunal) was set up later to resolve conflicts.

The period 2000–2010 saw the following developments:

In 2000, the government spun off a new entity, BSNL (Bharat Sanchar Nigam Limited) as a public-sector corporation, to provide Telecom services for the entire country, barring the two metros of Mumbai and Delhi already covered by MTNL. For cellular operators, the permissible stake that can be held by foreign companies (FDI—foreign direct investment) was raised to 74%. This boosted the coverage area and increased competition, thereby making Telecom services more affordable (lowest tariffs in the world), which in turn led to a greater number of mobile subscribers. Spectrum auctions were held in 2000 and 2001.

The state-owned long-distance international Telecom service provider, VSNL, was bundled off to the private corporate group, the Tatas, in 2008, being renamed as Tata Communications after the group took a 25% stake in VSNL. As this area had been declared open, all the Indian private Telecom service providers signed international roaming agreements with overseas operators.

From 2011 onward, there were radical shifts in the playing field.

The players that dominated the market are: Airtel, Reliance Communications, Jio, Idea, Tata Indicom, Vodafone, and BSNL. The fall in mobile phone tariffs led to mobile telephony becoming a stiff competitor to the landline, resulting in both becoming more efficient and innovative. In 2011, MNP (mobile number portability) was launched, which mandated that subscribers be allowed to carry their mobile number from one mobile service provider to another.

Due to the wide coverage across the length and breadth of India and its existing network, the incumbent player BSNL is a behemoth that still held sway over large parts of India, although the onrush of private players badly dented BSNL's profitability, causing it to make losses until 2013–2014. BSNL then strengthened its mobile footprint, increasing data capacity by 30 to 40 times, fixing gaps in 2G network, and expanding capacity to provide both 2G and 3G networks, by adding 27,000 towers (15,000 for 2G and 12,000 3G). It focused on improving its Wi-Fi, strengthening data network, and core network (OTN, MPLS, OFC), expansion of SGSN and GGSN, 3G/4G, and FTTH (fiber to the home). All these investments paid off and BSNL started becoming profitable from 2014–2015.

The spectrum auction in 2010 for 2G and 3G was very competitive and raised a lot of money, equivalent to US$17 billion in 2016 currency, over 183 rounds of bidding conducted in 34 days. All the top private players participated in the bidding for spectrum, and the public-sector corporations BSNL and MTNL were awarded spectrum without having to bid.

The 2012 auction was held for selling 2G band for GSM/CDMA. It was a disappointment for the government as it received bids of some US$1.5 billion, which was about a third of their expectations.

After that, spectrum auctions were held every year (2013, 2014, 2015, 2016) to progressively sell spectrum in different bands, 2G/3G/4G/5G. The auctions were not a complete success as the private players felt that the reserve prices were too high. Many refrained from bidding in some of the auctions. With a dozen players in the market, the competition intensified.

9.3   ICT Today in India

Here is a brief overview of the ICT in numbers, as of the year 2017.

Detail	Numbers
India's population	1.34 billion (Dec 2017) i.e. 1340 million
USA's population (for comparison)	325 million (Dec 2017)
Australia's population (for comparison)	24 million (Dec 2017)
Total phone subscriptions (December 2017)	1.207 billion
• Mobile users	1183 million
• Landline users	24 million
• Smartphone mobile users	400 million
smartphone penetration	30%
smartphone price	US$130
• Cost of data ranges from lowest (i.e. Jio's tariff) to industry average	$5 per 30 GB to $3 per GB
• ARPU (average revenue per user)	$2 per month
Total no. of PCs (personal computers)	150 million
PC penetration	~10%
Total Internet users (June 15, 2017)	~500 million
Internet penetration	34.4%
Internet penetration in urban India	60%
Mobile Internet users	159 million
Broadband (> 512 KBPS)	327 million
• Wired	20 million
• Wireless	307 million
Facebook users (June 15, 2017)	241 million
YouTube users	180 million
WhatsApp users	200 million
TV owning houses	170 million
DTH (direct to home) operators	6
Private TV channels (2015)	826

Sources: India Tech Online (http://www.indiatechonline.com/snapshot.php); Internet World Stats (http://www.internetworldstats.com/top20.htm)

As is clear from the above statistics, India is a mobile-first market, because mobile penetration is nearly 100%, whereas the PC penetration is only 10%. Within the mobile space, the smartphone penetration is 30%, and growing rapidly, with smart tablets also picking up. Evidently, the reasons for the strong skew towards mobile devices are affordability, portability (form factor of mobile), usability, besides the fact that most of the functions that people need/use are available through mobile apps and through web content customized for mobile browsers. Thus, it makes sense to launch more services on mobile devices to target the entire addressable population of India.

From 2015 mobile Telco market felt the heat of:

- High spectrum prices hurting their bottom line
- Severe competition in the market leading to price wars and fall in revenues
- Burden of upgrading the infrastructure to meet the ever-growing demands of the subscriber base, for example, providing higher bandwidth and services

Reliance, with its deep pockets, launched mobile services in September 2016 through its new company Reliance Jio Infocomm, providing services in all 22 Telecom circles of India, at rock-bottom prices, thereby acquiring 16 million subscribers within a month, and reaching 130 million subscribers by October 2017, that is, in a little over a year, it had captured 12.39% of the market share.

Consolidation in the market was inevitable. As of November 2017, the biggest mergers on the horizon are:

- Vodafone and Idea, with market shares of 17.68% and 16.20%, respectively
- Airtel, Tata Docomo, and Telenor, with market shares of 24.21%, 3.41%, and 3.81%, respectively

In March 2018, Aircel filed for bankruptcy, being unable to face the cut-throat competition in a market which saw call rates dropped to the lowest levels in the world, especially due to the aggressive entry of Reliance Jio. Aircel's Malaysian investor, Maxis, now stands to see its investment of US$7 billion being wiped out (Chew and Alexander 2018). Other players like Shyam Sistema have also exited the market.

This underscores the turbulence in today's Telecom landscape in India, despite India being the world's fastest growing economy, with youth-centric demographics and a billion plus population. Clearly new business models need to evolve, and the new regulations are intended to stimulate the sector. Currently due to protectionism (see chapter on India's regulations), the Telecom companies are making 50% of revenue from voice and 50% from data. However, when restrictions are removed and technologies like SIP Trunking are implemented, then the data usage should increase 40X by 2020 (E&Y 2016). This massive increase in data consumption would open up gigantic opportunities for the digital economy. It should be noted that India has 125 million people who speak English. Currently Tier 1 and 2 cities and towns (and not rural areas) dominate both the Internet and the consumption class, and the number of people who speak English. As the Internet penetration and the consumption class steadily grow into the rest of India, then 90% of the users will be non-English-speaking people. Regional language offerings will become both a challenge and an opportunity. India projects its digital economy to be about US$4 trillion by 2020 (*The Economic Times*, June 2017). The newly released draft National Digital Communications Policy 2018 takes most of these into account and is discussed in the next chapter on India's regulations.

9.4   THE ENTERPRISE LANDSCAPE

In the fiscal year 2016–2017, information technology and BPM (business process management) directly contributed to 7.7% of India's GDP (Statista 2018). As per NASSCOM, the sector-aggregated revenues in 2017 were about US$150 billion, the split-up being US$99 billion from exports, and US$48 billion from domestic revenues (NASSCOM 2017), growing at 7–8% annually. The IT-BPM sector is the largest employer in India, directly employing 3.7 million people (Make in India 2018). However, the bulk of these companies are IT services companies, not product companies. There are very few globally successful Indian product companies. This is a failure of the Indian IT industry. Even major digital companies that are run by software, for example, Flipkart and Paytm, have a large stake belonging to foreign companies like Softbank, Alibaba, and so on.

It is hoped that the new generation of product startups will take India to a leadership position on the world stage.

ICT has made tremendous progress in the past decade in India, with practically every sector seeking to adopt it. While it is impossible to do justice to this topic in a brief section, here are a few points as an indicator of things to come in every sector:

1. *Retail*: e-commerce stands at US$38.5 billion in 2017. The total online spending is growing at 31% per year. This includes both domestic and cross-border shopping of US$9 billion (IBEF 2018). Paradoxically, because India's physical retail industry is beset by poor infrastructure in terms of affordable rental space and capex, it is making the digital alternatives more attractive. Barring perhaps Tier 1 and 2 cities and towns, delivery logistics is still a major and unsolved problem for e-commerce owing to infrastructure problems, lack of proper addresses, high rate of rejection/return of goods, and labor problems. In India a package is handled by at least 20 people before it reaches the destination, whereas, as described elsewhere in the book, in the USA, Amazon needs only a minute of human labor to ship a package (Kola 2017).

2. *Tech Startups*: India is the third largest tech startup hub in the world, with 5400 startups as of 2018 (Make in India 2018).

3. *Health-Tech*: In the health sector, ICT is being used to address the basic problems in Indian healthcare, such as inaccessibility and insufficiency of quality care to the backward areas, lack of specialist doctors beyond the Tier 1/2 cities, improper capacity utilization of hospitals and resources, lack of patient information management system leading to fragmentation and duplication/bottom-up recreation of records diagnosis every time (Sharma 2018). The current developments across the board are the use of CRM (customer relationship management) and EPR (electronic patient records) in hospitals, alternate delivery model to increase reach such as remote diagnosis of patients, availability of a variety of medical insurances, automation in procurement and billing, increasing use of intelligent analytical and AI tools for diagnosis. Deployment of comprehensive HIMS (hospital information management systems), for example, in Apollo Hospitals. The government has an ambitious plan of providing health insurance cover to over ten crore families in India.

    A T-Hub startup Carengrow (www.carengrow.com), is a cloud-based preventive healthcare B2B2C solution, which monitors the health of school children. It uses data from screenings to do interventions and analysis of behavioral changes for providing guidance and corrective actions as regards their health.

Aravind Eye Hospitals (www.aravind.org) uses tele-medicine with mobile diagnostic centers so that doctors from the main hospital can diagnose and evaluate millions of patients to scale. Its research wing, Aurolab, has indigenously designed and developed an extremely low-cost lens for cataract patients.

4. *Manufacturing*: Using ERP, especially on the cloud, is becoming the norm. The next wave of adoption will see heavy adoption of IoT, big data analytics, and social media and mobility. The focus is on customer experience and co-creation and being able to run normal operations seamlessly. Practically all the manufacturing units are embracing digital transformation in some form or the other. For example, JSPL (Jindal Steel and Power Limited) has an IT center of excellence in Raigarh and Sonepat that are building a digital roadmap and are constantly innovating, adopting cutting-edge technologies, and implementing them in business (CIO 2017).

5. *Fin-Tech*: Banking and insurance are heavily digitalized now. Every single bank has gone online (cloud), has deployed ATMs, is accessible through mobile apps, and enables investments to be made online. Furthermore, banks are investing in modernization using the latest technologies such as big data analytics, blockchain, and AI. These are for a multitude of reasons such as fraud detection, analyzing customer behavior and tastes to provide superior user experience and get better ROI. Examples are provided in other chapters.

6. *Edu-Tech*: This sector is undergoing an unimaginable revolution with a plethora of tools, websites, and mobile apps. Using MOOCs (massive online open courses) is quite commonplace now. The entire cycle uses ICT heavily, right from education itself to getting jobs to improving career prospects through training/counseling/mentoring while in job.

Examples: Coursera/EdX/Udemy/Khan Academy/Open Yale/Academic Earth all serve excellent courses and content. Naukri/Monster/Shine provide job opportunities and connects. Portals like Skillenza allow students and professionals to participate in "hackathons" and get hired. One-to-one training is made possible through tools like Webex, Zoho Meeting, Skype, and so on.

9.5   Startups and Innovation

With the changing times, it is most appropriate to note how technological innovations are driving growth all over the world and bringing prosperity to all countries. Innovation is the new economy that keeps one competitive in the world. An entire nation's growth trajectory pivots on its ability to innovate around its own market's quirks.

With rapid influx of trends such as Artificial Intelligence, Machine Learning, and cryptocurrency, the health and future of an entity, be it a company or a country, is highly dependent on the intelligent adoption of today's changing technologies. There is a direct bearing on the economic growth and development.

India, at the cusp of its growth, is competing actively and changing the dynamics of the global economy. It took big strides in its startup ecosystem with the launch of the "Startup India" program in 2016. Indian startups are now diversifying across various economic platforms and have unleashed technological innovations to meet India's unique challenges. Startup India is also fostering entrepreneurship and has catapulted the country to be the world's third largest startup hub. In 2017, government initiated a few policies that also eased compliance requirements for foreign funding, recognizing the need for India's startup ecosystem to attract FDI. This, in turn, would create jobs and boost India's growth further.

Economically, India is growing faster than China and is set to be the third largest nation by 2022, and there are three factors that support this projection:

1. *Demographic Dividend*: According to the International Monetary Fund, 65% of the Indian population is below 35 years of age, and they contribute to a significant 2% of the GDP growth of the country. This is a massive competitive advantage when compared to developed nations like Germany, the USA, Switzerland, or China.
2. *Favorable Government Policies and Initiatives*: The environment is changing rapidly; the business-friendly outlook and initiatives of the current government are hoped to provide a fillip to India's economy to yield stability and growth. These include the regulations of ICT policies, Make in India, Startup India, and demonetization.
3. *Digital Revolution*: With the introduction of Aadhaar (digital identity), India Stack, and the underlying smartphone revolution, India projects a $4 trillion digital economy with 900 million users and 1 billion digital transactions. Besides, ICT is helping the government curb corruption, for example, linking Aadhar to pension schemes and LPG unearthed endemic scams.

It is projected that by 2020 there will be 11,500 firms, a rise from 5100 startups seen in 2016. In order to achieve this number, the government is making favorable amendments for tech-driven startups. Furthermore, the Indian market will change to accommodate this explosion. The infusion of VC and PE funding increased from US$13 million to US$1818 million between 2010 and 2014. Angel investment has also multiplied almost eight times from US$4.2 million to US$32.2 million. These trends are the indication of the times to come (Goyal 2017).

There are a number of factors that attract foreign investors to India, and the biggest factor is consumer growth, backed by the mobile revolution. Also, the focus of New York-based Tiger Global Management (TGM) has given confidence to other global private equity and hedge funds to come to India. Making big bets on Indian innovation has become a global point of interest. Adding to this list is global investors such as Alibaba and Softbank (among others).

The Startup India Standup India program launched by the current government is aimed at revolutionizing and accelerating the startup revolution in India. The Startup India program is explained more elaborately in another section. Indian startups are witnessing strong traction. In the past decade, we have seen the rebirth of the startup ecosystem with a more sustainable business model in the form of venture capital. This is now creating strong roots for centers of innovation, and we are seeing the results of this in cities such as Bengaluru, Hyderabad, and Gurgaon. For example, an online system for environment clearances, filing income tax returns, and extension of validity of industrial licenses to three years have been put in place. In addition, a few regulatory changes were made to attract global entrepreneurs to India:

1. The government increased foreign direct investment limits for most sectors resulting in a 62% increase in FDI equity inflows since the launch of Make in India.[1]
2. Protection of intellectual property rights of innovators and creators was ensured by upgrading infrastructure and using state-of-the-art technology.
3. The revolutionary GST (Goods and Services Taxes) has harmonized with the Indian market.

---

[1] Department of Industrial Policy and Promotion.

Perhaps as a result, India moved 30 notches higher in the World Bank rankings in ease of doing business and developing a conducive pro-business environment with good corporate governance practices for startups, and in the latest Global Innovation Index, India stands at No. 60, representing a growing culture of technological entrepreneurship, innovation, and a benchmark in R&D spending and investments.

9.6   POLICIES TO BUILD THE FUTURE OF INDIAN STARTUPS

2016 has been crucial for India, owing to two key financial decisions by the government in the past—demonetization and implementation of the Goods and Services Tax (GST). It included significant amendments intended to be in favor of startups, and for giving financial industry a healthy boost. On the heels of these developments in the policy environment, the year saw deeper penetration and deployment of newer technologies.

While continuing to harbor the third largest startup base in the world, India also jumped up 30 notches higher to make it into top 100 in the World Bank's "ease of doing business" rankings.[2] Needless to say, we still stand at the 100th position.

While most of the competing countries have created conducive pro-business environment, our government regulations for business license, excessive taxes, delayed judicial system, and the absence of strong domestic institutions with good corporate governance practices have still left us trailing behind at achieving a progressive startup ecosystem.

So, what does it take to become both an entrepreneurial and an economic superpower?

1. Swift starting up process with standard provisions for all types of businesses, except the restricted industry (Patnia 2018).
2. Uncomplicated and simpler licensing, land acquisitions, and construction-related approvals for manufacturing and hardware industry.
3. Faster and stricter enforcement of debt and other contracts, and quick turnaround of judicial remedies in case of disputes. A good IBC (Insolvency and Bankruptcy Code) will de-risk and encourage investment in startups (Ahluwalia 2018).

[2] Source: http://www.doingbusiness.org/data/exploreeconomies/india.

4. R&D spend for innovation (*The Economic Times* 2018a).
5. Tax exemptions for startups to grow and contribute towards GDP.

In addition to these measures, the need of the hour for Indian entrepreneurs is as follows.

9.7   ELIMINATING THE INSUFFICIENCY DIALOGUE

The government had introduced a $1.5 bn fund under its Startup India Action Plan. Two years later, only $92 mn of that amount has been sanctioned to a mere 74 startups out of 14,800 startups in India. Procuring these funds is a time-consuming ten-step process. Meanwhile, startup founders continue to use the phrase—"funding is hard in India". While there is a desperate need for capital access and distribution, startups that do raise money are being subject to confusing tax laws. If there is a certain mechanism for understanding and qualifying startups for the program, then there is certainty in building conducive environment for the growth of startups.

The angel tax is still wrapped in a haze of uncertainty with companies becoming liable for raising funds based on their evaluation rather than their fair market value. Currently, the angel tax rate stands at a whopping 30% which according to NASSCOM has resulted in a 53% drop in angel funding during the first half of 2017. Angel tax remains to be a bone of contention between authorities and investors, even after the government exempted "innovative" startups (Ghosh 2018). This was not removed even in the budget of 2018. The tax holiday period could also have been extended for the startup sector.

With about 5300 startups in the technology space, precious resources are being drained in responding to tax notices alone. Although it currently exempts startups that make less than Rs. 25 crores per year in its first five years, the angel tax certainly needs a 180-degree turn in perception so that the tax man can hold potential money launderers into account, rather than going after the entrepreneur.

Whether a startup is innovative or not currently depends on a certification by the Department of Industrial Policy and Promotion (DIPP), which is further weighed down by several other riders. This has led to several companies not being incubated inside government registered incubators or eligible for government grants.

A mechanism should be developed to provide this benefit to all genuine investors and invested companies. Also, the discretionary powers of the assessing officer should be removed. The valuation should either be process driven or based upon approval or suggestions from industry experts. The assessing officer taking approval from a more responsible commissioner-level senior could also be a way forward.

Explicit provisions must be made to empower these startups unconditionally. A quick turnaround to amend existing laws might actually be counterintuitive as it is bound to breed more confusion. What we need is a clear infrastructure that does not discredit the intent of building entrepreneurship in this country.

It would be interesting to see a few reforms on the ease of closing businesses. This can be addressed by fewer statutory compliance procedures.

9.8   The Biggest Payoff for Making in India

When we consider manufacturing, any new product that is invented in India and helps solve a socio-economic problem is eligible for subsidies for five years. However, there are products already in existence in other parts of the world that prevent Indian manufacturers from getting the necessary patent to avail that subsidy. Current infrastructure makes it extremely difficult for these products to be manufactured in India, especially if they don't have the required domestic partner.

The allowance for 100% FDI covers only verticals like single-brand retail. The healthcare sector is still left with no choice but to import expensive equipment and pay a steep import duty on it. Healthcare startups bear the conflict of existing in a country where healthcare is seemingly more economical, but the necessary infrastructure is not. The International Monetary Fund expects India's GDP to grow by 7.4% as against China's GDP that will only grow by 6.8%. This is not a mere sign of progress; it's also a big red flag that cries out for "policy innovation now", and not in the future. We cannot be our own external force that slows down this inertia of momentum.

It takes a bold move, such as exempting startups from income tax for the first two years, so that they can create job opportunities for others and experiment non-conservatively.

Hassle-free and incentive-based exports along with quick import of things should be addressed for the startups operating in hardware manufacturing sectors. Ideally it is beneficial for startups to have a harmonized code that reduces lead time in customs.

Government should invest in developing innovation zones with large-scale hardware ecosystem, manufacturing-related skill development, and development of the ecosystem with easy availability of designs, machinery, spare parts, and ancillary industries.

Another point is that Make in India need not mean manufacture in India (Aiyar 2015); it could mean design in India. Most of the high valued branded/premium goods of the developed countries are manufactured in China at a low cost. The bulk of the value is captured in the design, which lies in the developed country, and not in the manufacturing, that is, the manufacturer in China gets a tiny fraction of the price at which the premium product is sold in the market.

Hence, following the discussion in the chapter on winning the competition, Indian companies must embrace design thinking to become prosperous.

9.9   THE DIGITAL ECONOMY WE NEED

Though the Reserve Bank of India is already working towards bringing a regulatory framework to life, the Indian population has just started getting accustomed to mobile wallets and UPI. We neither have the knowledge nor the vision of how digital payments might create the future of our economy.

Several Indian entrepreneurs are forced to incorporate their companies in countries like Singapore and Estonia, and perhaps that explains why we must borrow policies that were already piloted in those countries.

According to industry reports, digital payments are projected to supersede cash by 2022. But, this remarkable momentum created in the digital payments ecosystem will need more policy support to be sustainable. The government has created crucial supply-side infrastructure—UPI, India Stack, eKYC, and Aadhaar—but it must address demand-side concerns and continue to incentivize digital payments and their providers to help the ecosystem gain greater adoption and synergy.

MAT should have been completely scrapped or minimized: Minimum alternate tax or MAT needs to be rationalized which is at the current rate of 18.5% (~21% with surcharge and cess), is very high, and significantly impacts cash flows of the companies which otherwise have low taxable income or have incurred losses in the past. With the phasing out of various exemptions, deductions, and reduced tax depreciation, there is a strong case for doing away with MAT or at least significant reduction in MAT rate, so that it retains its true character as alternate "minimum" tax (*The Economic Times* 2018b).

In the budget of 2018, the finance minister has emphasized the importance of latest technologies such as robotics, AI, digital manufacturing, big data analysis, and Internet of Things (IoT). He doubled the allocation on the Digital India Program to INR 3073 crores. The budget also declared that the planning body, NITI Aayog, will establish a national program for AI (artificial intelligence), which will drive government's efforts in AI for national development (Thomas 2018). Other countries like UAE have also recognized the importance of AI, as is evident from their appointment of a Minister of AI (Arabian Business 2017).

There is definitely hope, and it can only be done by speeding up and optimizing existing policies, introducing a large and tangible payoff for startups in India, and innovating for 2025 right away. As the world's fastest growing economy, the only thing we need to stay ahead of is ourselves. The National Digital Communications Policy 2018, a draft policy released by the Department of Telecom on May 1, 2018, addresses many of these areas and is discussed in the next chapter.

9.10   NAVIC (INDIAN REGIONAL NAVIGATION SATELLITE SYSTEM)

To remove dependence on the American GPS (Global Positioning System), ISRO (Indian Space Research Organisation) launched satellites for the project, NAVIC (Navigation with Indian Constellation). The system of satellites will provide real-time positioning and timing services, with a position accuracy of 10 meters over the Indian landmass and 20 meters over the Indian Ocean. The coverage will also extend to 2000 km around India.

Why the GPS cannot be relied upon:

- During the hostile intrusion by Pakistan into Indian territory in 1999, the USA refused to share military GPS data with India which would have helped Indian troop movements in the difficult Himalayan terrain.
- During the second US-Iraq war, the GPS time and position details were readjusted (tampered with) to mess up the adversaries of the USA.

Thus, India embarked on the NAVIC project with the goal of achieving self-sufficiency. NAVIC currently has seven satellites in orbit, with two on the ground, in a stand-by mode. It was planned for commercial launch in

2018; it suffered setbacks when IRNSS-1A failed, as also the mission to replace this satellite with another in 2017.

The data it provides currently is quite coarse, but the Indian government has approved increasing the satellite constellation from 7 to 11. NAVIC is intended to work in two levels: standard one, for civilian use, and restricted (encrypted for military and other special uses requiring government authorization).

Unlike GPS, NAVIC's coverage is not global, but regional. So, India is working on GINS (Global Indian Navigation Satellite), which is a different and futuristic program that will have better accuracy than GPS (which uses older sensors). GINS will be a constellation of over 24 satellites orbiting the earth at a distance of 24,000 km.

For civil aviation, India has another satellite system that is GPS aided, called GAGAN (Geo Augmented Navigation System). The goal of GAGAN is to provide GPS-based navigation cues for all phases of an aircraft's flight over the Indian airspace and surrounding areas. It meets the standards of international civil aviation regulatory bodies. GAGAN is a regional navigation system, not a global one.

GPS has two scales: military and civil, which have an accuracy of one meter and three meters, respectively. As India does not have access to military GPS, it uses civilian GPS for course correction before other guidance systems are used. For catering to its military needs, India may integrate the regional systems with GLONASS (Russian equivalent of GPS).

Most smartphones have about five inbuilt antennae to cater to the different bands. It is not hard to add an extra antenna. In the future, for making NAVIC available to the Indian population, India's regulations may mandate that all cellphones being released in India be equipped with the antenna that can receive NAVIC signals.

With NAVIC and other systems, India can make rapid strides with innovations in practically every area. It empowers everyone with real-time and precise information about place and time.

Use-cases include:

- Navigation of vehicles, aircrafts, and ships
- E-commerce deliveries and location-based services for mobile phones/devices
- Disaster management and timely relief/aid, emergency health services
- Border mapping, surveys of lands/forests/rivers/and so on
- 3D maps by combining with remote sensing

- Power grids synchronization
- Monitoring buildings and other structures
- Monitoring all movements on land/air/sea/rivers of people and goods, monitoring logistics and helping in infrastructure projects, fishing, mining, manufacturing
- Banking and finance: money transfers, tracking assets, and so on
- Aviation, travel, tourism
- For military and police: missile guidance, crowd control, counter-terrorism, anti-crime measures, tracking people, equipment, and goods/artifacts of interest

9.11   OTT (Over-the-Top) Services

Telcos have run into a major problem worldwide, and especially in India. The OTT players have not only cannibalized the traditional voice and messaging services but are also capturing the market for all the new business models such as LBS (location-based services) and IoT. The Telco operators in India have been hit hard because India's infrastructure is still in its early stages and heavy investments need to be done. And the burden of these investments falls largely on the Telcos, not the OTT players. Whereas in the developed Western countries a lot of investment was done decades ago in backbone and access networks, and so the Telcos there have recovered their monies over time, in India there seems to be no ostensible way for Telcos to get a return on investment for making heavily in the infrastructure when the OTT players happily skim off most of the profits. Prices of Telecom services such as voice, messaging, and broadband are in a free fall mode for both landline and mobile telephony, and consumers are demanding even more while paying next to nothing. As the content grows richer, such as high-definition online video streaming, and demand increases steeply, the pressure on the infrastructure goes up. With falling prices, the Telcos stand to make losses and even engage in price wars with one another.

As per Creaner (2017), Telcos need to undertake digital transformation to improve the customer experience for both enterprises and retail customers. The Telco's transformation should improve the speed and efficiency of its business and help maximize revenue from both traditional and new services.

Telcos can find several ways out of losing the battle to the OTT players:

1. Partnering with the OTT players by providing them with valuable data such as consumer behavior, analytics, and insights gained using AI/Machine Learning.
2. Taking the lead in innovation and offering newer business models themselves.
3. Partnering with governments and enterprises such as energy, utility, transport, agriculture, healthcare companies to provide them with comprehensive and end-to-end solutions in different areas such as IoT, relevant content and analytics/insights, putting up secure/scalable websites, content management, and so on.
4. Selling superior broadband services in a tiered manner to enterprise customers.
5. Acquiring or tying up with startups in different domains in the right, relevant areas as determined by innovative and profitable new business models.
6. Providing specialized services at a premium, such as emergency services, piping reliable Telecom services into malls and buildings to provide superior service, addressing safety issues such as providing good connectivity sans high radiation, special services for travelers and commuters.
7. Providing value-added services to customers, such as cloud services, storage, secure encrypted services, and so on.
8. Selling devices for different customer segments through package deals, for example, mobile phones, fixed line phones, Wi-Fi hubs, switches, routers, storage devices, connectors, and so on.
9. Collaborating with or acquiring companies that are into media, advertising, entertainment, content development, software development, IT services, contact center and support, for providing richer, innovative services.
10. Investing in or acquiring innovative startups.

9.12   UPI (Unified Payments Interface)

The NCPI (National Payments Corporation of India) launched UPI, which enables people to transfer money instantly to other people or to make payments to vendors for purchases (Cashless India 2016). This can be done through smartphones. This is a major move towards demonetization, and to propel a digital economy, that is, the Digital India Initiative.

UPI is like an email ID for payments. It can do away with credit cards, debit cards, electronic wallets, and other cumbersome payment systems like NEFT and RTGS. With UPI, bank account holders can send/receive money from their smartphones using just their:

- Aadhar number as unique identity
- Mobile number
- Virtual payments address

There is no need to enter any bank account number, bank name, IFSC code, and so on or set up payee details for the bank to validate (which is not instant). Only banks (and not mobile wallets or Telco operators) are allowed to become the payment service providers of UPI as per RBI's directive. There will be a factor authentication called MPIN, which will have to be entered to validate the transaction. UPI is a great boost to the Indian banks. As it matures, more sophistication will be added, for example, detailed transaction reporting and resolution of disputes as regards payments or delivery of goods and services. UPI is in fact the upgraded version of IMPS (immediate payments service), which already caters to about Rs. 8000 crores of transactions per day. Hence there is already a precedent and a solid base for UPI to take off. It will revolutionize Indian banking.

9.13   India Stack

The India Stack is a software stack based on open-source software technologies and is about the largest open API (application programming interface) in the world today. It's built by the group iSPIRT, a non-profit organization that seeks to boost software product development in India. The India Stack is a common, re-usable software platform that attempts to solve the major problem, specific to India's digital initiatives, that is, the lack of an underlying integrated digital infrastructure that provides a minimum subset of essential functions:

- Identity management, based on an accepted and government-approved identity system
- Executing and managing transactions
- Digital assets management that allows access to digital systems with billions of artifacts

- Electronic payment system that is cheap and compliant to regulations
- Secure data sharing, especially for private data

Governments, businesses of all kinds, large and small in India, face an unsurmountable challenge when they wish to build a digital service in India. They need a pre-built, ready-to-use, integrated, and scalable infrastructure stack on which they can build their digital service.

Why the need for India Stack? To improve the public distribution system, access to services, and for transparency. Apart from this, this also opens business opportunities on a national scale. There is a dire need to unlock the potential of the next 400 million people, as discussed in this chapter's next section, Skill India.

About 160 million people receive PDS (public distribution system) benefits, and the public healthcare spending is 1.4% of India's GDP. There are rampant inefficiencies and corruption everywhere in the system.

Cases in point are the mammoth frauds exposed by a simple application of ICT, the mere linking of the Aadhar ID with each of the following:

- LPG Ration cards: Exposed 50 million ghost accounts and 35 million fake LPG connections (Aadhar LPG Scam)
- Ration Cards: Exposed 23 million fake ration cards (Aadhar Scams)
- Pension (Aadhar Pension Scams)
- Mid-day Meal Scheme: Exposed 4.4 Lakh ghost students across three states (Aadhar Meal Scams)

The India Stack has four technology layers (India Stack 2018):

1. *Consent Layer*: This is the topmost layer. It provides a state-of-the-art privacy data sharing framework, open personal data store. Thus it enables data to move securely and freely to level the playing field for markets' access to data.
2. *Cashless Layer*: This provides transition to cashless economy through game-changing electronic payment systems, such as UPI, NEFT, IMPS, AEPS, APB. It provides a single interface to all the country's bank accounts and wallets to facilitate easy payments.
3. *Paperless Layer*: This layer provides access to an exploding base of paperless system with billions of artifacts, for example, Aadhar, eKYC, Digital Locker, E-Sign. It ties digital records to an individual's

identity, thereby making redundant the massive paperwork that is otherwise required for collection and storage.

4. *Presenceless Layer*: This is the bottom-most layer where a universal biometric digital identity facilitates people to participate in any service from anywhere in the country. It provides a unique digital biometric identity with open access of nearly a billion users. This layer primarily uses JAM (Jan Dhan, Aadhar, and Mobile).

The India Stack is supported by three invisible pillars:

- *Pillar 1: Jan Dhan Yojana*—This is a national mission for financial inclusion through lowering the entry barriers to banking for the Indian masses, especially those at the bottom-of-the-pyramid. This is to ensure that the entire Indian population has easy and affordable access to financial services, such as banking, savings, deposit accounts, remittance, credit, insurance, pension, and so on. By June 2017, there were about 300 million accounts with deposits worth Rs. 698.4 billion, that is, over US$10 billion (PMJDY). The India Stack provides access to and enables transactions for this massive base of bank accounts. As per the scheme, the account holders have RuPay debit cards, accidental insurance cover up to Rs. 100,000. The biggest advantage is that NPCI (National Payments Corporation of India) has facilitated funds transfer and balance checking not only through mobile smart phones but also through normal mobile phones.

  The NUUP (National Unified USSD Platform) has provided integration of banks and mobile companies to enable mobile banking for the population at the bottom-of-the-pyramid. This is a major step towards financial inclusion.

- *Pillar 2: Aadhar*—This is a unique and government-authenticated identity for all citizens of India by the UIDAI (Unique Identification Authority of India). It includes biometric authentication and provides a unique number as an ID. Aadhar is being linked to all bank accounts, financial investments, and even public distribution services such as cooking gas, electricity, and so on. The is an accepted proof of identity everywhere in India.

- *Pillar 3: Mobile*—India has over 650 million mobile users by October 2017, of which over 300 million have a smartphone according to Counterpoint Research. Hence India is second only to China in the world, as regards size of the smartphone market. The number of

Internet users is about 450–465 million, as per IMAI (Internet and Mobile Association of India). Hence India is a mobile-first market.

The overall Internet penetration is around 31%. However e-commerce as a percentage of total retail sales is a mere 1.7%, as compared to China (19%) or the UK (17%). The positive indication is that the retail e-commerce growth is expected to be about 23% year-on-year (Statista 2018); the main reason is of course that India is a mobile-first market.

India's indigenous digital payments app, BHIM (Bharat Interface for Money) launched by the NPCI, enables fast and secure cashless transactions using mobile phones and is live in 59 banks. It was launched in Dec 2016, and by Oct 2017 it has been downloaded 20 million times and is seeing 7.63 million transactions per month with monetary value of Rs. 2349 crores (US$361 million). Its competitors, the apps Tez from Google and PhonePe from Flipkart, also have an overwhelming response. Google claims Tez registered 30 million transactions in the five weeks since it was launched in Sept 2017.

9.14   Skill India

In July 2015, the then newly elected Indian government announced the Skill India program. Considering India's workforce of 500 million people, the scheme initially targeted skilling 400 million people by 2022 (Skill India). Thus, this program was considered so important that the government formed a new Ministry of Skill Development and Entrepreneurship, MDSE. However, by June 2017, in two years, it managed to train only about 12 million people, thereby falling short of the target by a wide margin. It is evident that urgent and extreme measures are needed to bridge the yawning gap. As described earlier, ICT is a powerful multiplier, and can rescue the situation. The government has stepped up its efforts and is leveraging ICT in a big way. The MDSE is adopting the following strategies:

1. *MOOC (Massive Online Open Courseware)*: As mentioned earlier, there is a dire need to disseminate knowledge and skills on a national scale. The channels that can reach the population are Internet through mobile devices, television through satellite and cable, broadband Internet, and radio. MOOCs through the Internet are the top medium of choice.

2. *Building Entrepreneurs*: The different schemes are creating entrepreneurship hubs; providing entrepreneurship curriculum in both online and mainstream courses; networking entrepreneurs with peers, mentors, and incubators; fostering an entrepreneurship culture; promoting women entrepreneurs and those from underrepresented groups; encouraging social entrepreneurship and grassroots innovation.

3. *National Skill Development Mission*: The key building blocks are institutional training, infrastructure, convergence, trainers, overseas employment, sustainable livelihoods, and leveraging public infrastructure (National Skill Development Mission 2018).

4. *Establishing Standards and Quality*: Under this, the NSQF (National Skills Qualification Framework) organizes competencies into a framework based on grade and levels based on knowledge, skills, and aptitude. This is much needed to provide a quality workforce and pave the way for individual growth. The NOS (National Occupational Standards) provide specifications for standards of performance/skills against a specific activity, and QPs (Qualification Packs) are a set of NOSs that map to a job role, and these are done for all job roles in the industry.

5. *Training Providers*: With the view that they are very important players in the ecosystem, the training providers will be provided every help like financial incentives, for example, loans, grants, tax breaks, and equity. And they would be evaluated on several criteria to ensure that they are capable of delivering the most value to the industry.

6. *Innovations in Skills Development Marketplace*: The National Skill Development Corporation (NSDC) has set up forums and committees to invite and review and host innovations pertaining to skills development coming from all over the country. This would boost ideas, solutions, models, pilots and practices related to skill development. Setting up knowledge centers, hosting innovation challenges, and attracting investment are part of its strategy.

7. *World Skills*: With this initiative the NDSC wants to promulgate vocational skills and champion learning for contribution to the society at a level on par with international standards. It invites sponsorships and holds world skills competitions for various skills.

8. *Other Schemes*: Various other schemes were launched by the Indian government to address specific needs and/or specific groups and provide incentives. PMKVY (Pradhan Mantri Kaushal Vikas Yojana) for four years (2016–2020) to benefit ten million youth, Udaan for Jammu and Kashmir region, polytechnic schemes, and vocationalization of education.

9.15   DIGITAL INDIA INITIATIVE

Launched by the Indian government in July 2015, this initiative aims to bring high speed Internet connectivity to rural India, which is home to 70% of India's population (2011 Census Data). The new National Digital Communications Policy 2018 also takes this initiative further and is discussed in the next chapter.

The Digital India Initiative has nine pillars:

1. *Broadband Highways*: This has three branches, namely, to provide broadband for:
   (a) All Urban: Leverage VNOs (virtual network operators) for coverage and all urban infrastructure to include communications
   (b) All Rural: Covering 2.5 Lakh village panchayats, with the department in charge being the DoT (Department of Telecommunications)
   (c) All National: Integrate national networks like SWAN, NKN (National Knowledge Network), and NOFN (National Optical Fiber Network) and enable national and state data centers with cloud. MeitY (Ministry of Electronics and Information Technology) would be in charge.

2. *Universal Access to Mobile connectivity*: To increase mobile penetration to rural areas covering 42,300 villages. DoT would be the nodal department.

3. *Public Internet Access Program*—which has two components:
   (a) CSCs (common service centers): MeitY would run this program. Scale up the reach to all the 250,000 gram panchayats, in order to deliver government and business services.
   (b) Post offices to now function as multi-service centers, managed by the Department of Posts.

4. *E-Governance*: Reforming government through technology. This involves simplifying government processes through re-engineering and using IT for automation and efficiency. To achieve this, several guiding principles and means are to be employed. Guiding principles constitute forms simplification, online applications, online repositories, and integration of platforms and services. Besides these, the other means mentioned are electronic databases, workflow automation, and feedback through public grievance redressal (which identifies areas where process improvements are much needed).

5. *E-Kranti*: Electronic delivery of services. These include 31 mission mode projects for e-governance, in the following areas:

e-Education, e-Healthcare, technology for farmers, technology for security, technology for cybersecurity, technology for financial inclusion, technology for justice, technology for planning

6. *Information for All*: Online hosting of documents and information and open data platform easily accessible by the people, government interactions with citizens using social media, web platforms, and online messaging.
7. *Electronics Manufacturing*: The target is that of imports-exports balancing out to net zero. This requires changes in many areas ranging from taxes to development of incubators. There are focus areas such as setting up FABS, fab-less designs, mobiles, electronics, smart energy meters, and so on.
8. *IT for Jobs*: To score high improvement in employment, IT can be used for job creation. DEITy is the nodal department for this scheme. It plans many large-scale trainings at different levels. For example, ten million students are to be trained in IT for jobs, setting up of BPO trainings, and so on.
9. *Early Harvest Programs*: This program attempts to seize clear opportunities for using several available technologies. For example, mass messaging, biometric attendance for all government employees, Wi-Fi in all universities, SMS-based weather alerts including disaster warnings, e-books as school books, and so on.

9.16   STARTUP INDIA INITIATIVE

Technology startups are also driving ICT growth in India; there are about 4700 registered startups (Make in India 2018). This initiative was launched by the government of India, under the Ministry of Commerce and Industry to boost innovation and growth of startup companies in India and generate large employment opportunities as the startups scale. DIPP (Department of Industrial Policy and Promotion) defines a startup as a company that meets these three criteria:

- *Not older than seven years. Relaxed to ten years for biotechnology startups*
- *With annual turnover not exceeding INR 25 crore (USD0.38 million) in any preceding financial year*

- *Working towards innovation, development, or improvement of products or processes or services, or if it is a scalable business model with a high potential of employment generation or wealth creation*

The Startup India Initiative aims to help startups by simplifying processes, handholding them in much needed areas such as assisting with IPR legal services, providing tax breaks to both investors and startups, helping startups get funding, and building incubators and ecosystems to nurture startups.

There are three pillars for the Startup India's action plan:

1. *Simplification and Handholding for Startups:*
   (a) Self-certification to keep compliance cost and burden low.
   (b) Startup India hub—one-stop-shop for the startup ecosystem and to access knowledge and investment.
   (c) Mobile app and portal—for startups to register and collaborate in the ecosystem.
   (d) Inexpensive and speedy legal support and patent examination—facilitate IPR (intellectual property rights) services and SIPP (startup intellectual property rights protection).
   (e) Facilitate participation of startups in public procurement through relaxed norms, to provide them level playing ground with established companies.
   (f) Faster exit for startups—help startups close business quickly if their business doesn't work out.

2. *Incentives and Support for Funding for Startups:*
   (a) Providing funding support through a fund of funds with a corpus of INR 10,000 crores—to facilitate a venture fund to kick-start the startups.
   (b) Credit guarantee fund for startups—to provide a convenient debt-financing scheme through traditional banking in order to facilitate credit for disruptive startups.
   (c) Tax break on capital gains—to funnel the money gotten from sale of capital assets as investments into startups.
   (d) Tax break to startups for three years—to help startups with their working capital needs.
   (e) Tax break on investments above fair market value—for raising seed-capital for startups.

3. *Industry-Academia Partnership and Incubation for Startups:*
   (a) Startup fests for showcasing innovation and providing a collaboration platform—to provide national and global visibility to Indian startups and spur the startup ecosystem.
   (b) Atal Innovation Mission (AIM) program with Self-Employment and Talent Utilization (SETU)—to provide world-class innovation hubs, forums, and events to boost startups.
   (c) Drawing on private sector expertise for incubator setup—to set up incubators across India in a public-private partnership mode and frame a policy framework for the success of such incubators.
   (d) Innovation centers at national institutes—to advance innovation by collaboration of incubators and R&D centers.
   (e) Seven new research parks modeled on the research park set up at IIT Madras—to promote innovation by fostering collaboration between academia and industry through joint R&D programs and incubation.
   (f) Startups in the biotechnology sector—for promoting entrepreneurship in this field.
   (g) Innovation-focused programs for students—to propel innovation culture among science and engineering students.
   (h) Annual incubator grand challenge—for incubators in India to match world-class standards.

The best example is T-Hub (Technology Hub), the immensely successful startup growth engine in Hyderabad, Telangana. It is India's largest and fastest growing startup engine, catalyzing innovation, scale, and deal-flow. For the startups, T-Hub provides incubation and access to mentors, investors, and academia.

9.17   ICT FUTURE SKILLS

The digital economy will create 30 million employment opportunities by 2024–2025 (MeitY and McKinsey 2018).

India's National Association of Software and Services Companies (NASSCOM) conducted a study with BCG (Boston Consulting Group) and has identified eight technologies that will grow exponentially and drive the industry in the future, both globally and nationally. It also identified 55 job roles for the future and the skills needed for these roles. The eight technologies are:

AI (Artificial Intelligence), VR (Virtual Reality), robotic process automation, Internet of Things (IoT), big data analytics, 3D printing, cloud computing, social and mobile.

In the WCIT (World Congress on IT) on February 19, 2018, NASSCOM launched a platform called Future Skills to help upskill two million ICT professionals and also another two million students and future ICT employees in the coming few years (Mendonca 2018). This platform is intended to be a one-stop-shop for all learning requirements for the future job roles and skills. As discussed in the previous chapters, ICT will pervade all sectors; hence the employees in those sectors will also need to learn relevant ICT technologies. NASSCOM signed an MoU with MeitY (Ministry of Electronics and Information Technology) for the initiative on providing skilling for the total spectrum of identified skills. The Future Skills initiative will complement the government's Digital India Program.

Individuals can sign up on the platform and track their progress and get counseling and career guidance for their career paths. This will help the individuals, organizations, and the country to stay relevant and create value and jobs, in the face of rampant automation which will remove many jobs as discussed in the earlier chapters.

9.18   Conclusions

India is one of the oldest civilizations on earth, and it has made enormous contributions in every field of human endeavor, including technology. India is among the largest economies in the world and certainly the most diverse. India's history with ICT and its current state in ICT are discussed, and how India is taking rapid strides in ICT to help jumpstart its economy, society, and its industry. ICT can be the real lever to help India become a superpower. A study of Indian ICT can help every economy, organization, and individual become successful in many ways, because there are many parallels across the world.

The Takeaway Box: With Its Youth, Talent, and Rich Heritage, India Will Be a Superpower by Leveraging ICT

- India is the fastest growing economy in the world and has 65% of its population below the age of 35 and 50% of its population below the age of 25.
- About 70% of India's working population works in the unorganized/informal sector, that is, they are not employed in formal jobs. This means that India is already well prepared for the "gig economy" that is coming up in the digital economy.
- It is the third largest economy in the world in PPP terms.
- India is making gigantic strides to move towards a digital economy, solve its problems, and become a real growth engine.
- It will rise both internally (India's domestic markets) and internationally (foreign trade) to take the center stage in the world.
- With its rich cultural and spiritual heritage, India has so much to offer to the world.
- ICT will help India leapfrog and even pole-vault to the position of a superpower.
- Collaborating with or investing in India in any form is the best thing to do today.

REFERENCES

Aadhar LPG Scam. (n.d.). Retrieved from http://www.business-standard.com/article/current-affairs/aadhaar-exposes-5-cr-ghost-accounts-3-5-cr-fake-lpg-connections-goyal-117110400648_1.html.

Aadhar Mid-day Meal Scam. (n.d.). Retrieved from http://www.hindustantimes.com/india-news/midday-meal-scheme-aadhaar-exposes-4-4-lakh-ghost-students-across-3-states/story-9Kna9AHXQYX59bn4bzvrzO.html.

Aadhar Pension Scam. (n.d.).

Aadhar Scams. (n.d.). Retrieved from http://www.thehindu.com/news/national/aadhaar-helped-save-about-50000-crore-through-dbt-centre-tells-supreme-court/article18937378.ece.

Aggarwal, V. (2018). *Sir Jagadish Chandra Bose—The Unsung Hero of Radio Communication*, MIT. Retrieved from http://web.mit.edu/varun_ag/www/bose.html.

Ahluwalia, M. S. (2018). *A Policy Wish List for the New Year, Live Mint*. Retrieved from http://www.livemint.com/Opinion/XIbWCVImQ2ShUjA8XY0HiO/A-policy-wish-list-for-the-new-year.html.

Aiyar, S. (2015, May 31). *The Times of India*. Retrieved from https://blogs.time-sofindia.indiatimes.com/Swaminomics/make-in-india-and-manufacture-in-india-are-two-different-things/.

Arabian Business. (2017, October 22). *UAE Appoints Minister for Artificial Intelligence*. Retrieved from http://www.arabianbusiness.com/politics-economics/381648-uae-appoints-first-minister-for-artificial-intelligence.

AYUSH, Ministry of AYUSH, India. (n.d.). Retrieved from http://ayush.gov.in/.

Cashless India. (2016). Unified Payments Interface. Retrieved from http://cash-lessindia.gov.in/upi.html.

Chew, E., & Alexander, S. G. (2018, March 2). Tycoon's $7 Billion Wipeout Turns His India Dream Into Nightmare. *Bloomberg Business Week*. Retrieved from https://www.bloomberg.com/news/articles/2018-03-01/tycoon-s-dream-of-india-riches-turns-into-a-7-billion-nightmare.

CIO. (2017, February). Indian Manufacturing Industry En Route to a Shining Future. Retrieved from www.cio.in/cio-interview/indian-manufacturing-en-route-shining-2017.

Creaner, M. (2017, September). Telco Digital Transformation: The Conditions, Journeys, and Destinations. *Huawei*. Retrieved from http://www.huawei.com/en/about-huawei/publications/winwin-magazine/plus-intelligence/telco-digital-transformation.

Das, G. (2012, September). *India Grows at Night*. Penguin India. ISBN-10: 0670084700, ISBN-13: 978–0670084708.

E&Y. (2016). Future of Digital Content Consumption in India, EY. Retrieved from www.ey.com/Publication/vwLUAssets/ey-future-of-digital-january-2016/$FILE/ey-future-of-digital-january-2016.pdf.

Economic Times. (2017, June). Digital Economy Can Reach $4 Trillion in 4 Years: Tech Sector to Government. Retrieved from https://economictimes.indiatimes.com/news/economy/indicators/digital-economy-can-reach-4-trillion-in-4-years-tech-sector-to-government/articleshow/59188885.cms.

Focus Economics. (2018). The World's Largest Economies. Retrieved from https://www.focus-economics.com/blog/the-largest-economies-in-the-world.

Ghosh, D. (2018, January). Budget 2018: Startup Investors Seek Abolition of Angel Tax.

Goyal, M. (2017, August 29). Why India Is Seeing a Fresh Wave of Global Innovation Centres, and How It Could Be a Lifesaver for IT Firms. *Economic Times*. Retrieved from https://economictimes.indiatimes.com/tech/ites/why-india-is-seeing-a-fresh-wave-of-global-innovation-centres/article-show/60238228.cms.

IBEF. (2018). E-Commerce Industry in India. Retrieved from www.ibef.org/industry/ecommerce.aspx.

India Stack. (2018). Retrieved from www.indiastack.org.

Kola, Vani. (2017). *India Macro and Consumption, Kalaari Capital.* Retrieved from https://blog.kstart.in/india-macro-and-consumption-a8615f7f9110.

LAWS, Lethal Autonomous Weapon System. (n.d.). Retrieved from https://en.wikipedia.org/wiki/Lethal_autonomous_weapon.

Make in India. (2018). *IT-BPM Sector Survey.* Retrieved from www.makeinindia.com/article/-/v/sector-survey-it-bpm.

MeitY and McKinsey. (2018, February). *India's Trillion Dollar Opportunity,* Ministry of Electronincs and Information Technology. Retrieved from http://digitalindia.gov.in/writereaddata/files/1.%20Trillion%20Dollar%20Economy.pdf.

Mendonca, Jochelle. (2018, February 19). *Economic Times.* Retrieved from https://economictimes.indiatimes.com/small-biz/startups/newsbuzz/narendra-modi-launches-nasscoms-futureskills-platform/articleshow/62982879.cms.

Money Control. (n.d.). Retrieved from http://www.moneycontrol.com/news/business/startup/taxing-the-angel-startup-investors-seek-abolition-of-angel-tax-in-budget-2018-19-2477945.html.

NASSCOM. (2017). *The IT-BPM Sector in India—Strategic Review 2017.* Retrieved from www.nasscom.in/knowledge-center/publications/it-bpm-industry-india-2017-strategic-review.

National Skill Development Mission. (2018). Retrieved from http://www.skilldevelopment.gov.in/nationalskillmission.html.

NDSC, National Skill Development Corporation. (n.d.).Retrieved from http://innovation.nsdcindia.org.

Panchal, Salil. (2018, January 31). IMF India growth forecast achievable, say economists, Forbes India. Retrieved from http://www.forbesindia.com/article/leaderboard/imf-india-growth-forecast-achievable-say-economists/49241/1.

Patnia, Alok. (2018, January). Things Budget 2018 must address for Startups in India, Your Story. Retrieved from https://yourstory.com/2018/01/budget-2018-startups-india/.

PMJDY, Pradhan Mantri Jan Dhan Yojana. (n.d.). Retrieved from https://www.pmjdy.gov.in/account.

Russell, Stuart. (n.d.). Retrieved from www.autonomousweapons.org.

Sen, Amartya. (1981). *Poverty and Famines: An Essay on Entitlement and Deprivation.* Oxford University Press. p. 39. ISBN: 978-0-19-828463-5.

Sharma, Mihir. (2017, September 7). India's burgeoning youth are the world's future, *Live Mint.* Retrieved from https://www.livemint.com/Opinion/2WSy5ZGR9ZO3KLDMGiJq2J/Indias-burgeoning-youth-are-the-worlds-future.html.

Sharma, Ajay Shankar. (2018). Healthcare IT in India, Asian Hospital and Healthcare Management. Retrieved from www.asianhhm.com/information-technology/ajaysharma-interview.

Skill India. (n.d.). Retrieved from www.skilldevelopment.gov.in.

Startup India. (n.d.). Retrieved from https://www.startupindia.gov.in/.

Statista. (2018). IT-BPM sector in India as a share of India's GDP. Retrieved from www.statista.com/statistics/320776/contribution-of-indian-it-industry-to-india-s-gdp.

The Economic Times. (2018a). Finmin Likely to Tweak Mat Norms To Boost Industry. Retrieved from https://economictimes.indiatimes.com/news/economy/policy/finmin-likely-to-tweak-mat-norms-to-boost-industry/articleshow/62511016.cms.

The Economic Times. (2018b). How startup lessons from Israel can turn India into an innovation economy. Retrieved from https://economictimes.indiatimes.com/small-biz/startups/features/how-startups-lessons-from-israel-can-turn-india-to-an-innovation-economy/articleshow/62545144.cms.

Thomas, A. (2018, February). Budget 2018 for Startups: Yea, Nay or Somewhere in Between?, *Economic Times*. Retrieved from https://economictimes.indiatimes.com/small-biz/startups/newsbuzz/budget-2018-for-startups-yea-nay-or-somewhere-in-between/articleshow/62741257.cms.

UN Urban Habitat Reports, UN Habitat's online reports, UN Habitat. (n.d.). Retrieved from http://urbandata.unhabitat.org/compare-cities/.

Vatsa, A. (2012, November 18). When Telegraph Saved the Empire. Indian Express. Retrieved from http://archive.indianexpress.com/news/when-telegraph-saved-the-empire/1032618/.

Wikipedia, Demographics of India. (n.d.). Retrieved from https://en.wikipedia.org/wiki/Demographics_of_India.

Wikipedia, Economy of India. (n.d.). Retrieved from https://en.wikipedia.org/wiki/Economy_of_India.

## CHAPTER 10

India's Regulatory Environment and Response to International Trade Issues

Yatha Raja Tatha Praja. As the King, So the People.
—*Sanskrit Saying*

10.1   INTRODUCTION

India has woken up to the new realities of the current day and is aggressively pursuing transformation on several fronts, as was highlighted in the earlier chapter. For decades, the kind of multi-party democratic system framed in its constitution, the electoral process, the bureaucracy, and the procedures for passing laws had rendered it quite slow in decision-making and execution. The laws passed had to appeal to a large variety of electoral constituents, and had to be driven by consensus, thereby forcing compromises in several areas. It is true that India has made great strides on several fronts, as described in the earlier chapter, such as space, defense, nuclear, key civil works, and reputed institutes of learning.

However, considering the development of India's populace as a whole, for nearly half a century since its independence, India pursued a socialist type of agenda and missed out on several chances for modernization and thus fell far behind on the world stage. Two such instances are privatization and mobile telephony, which could have been brought in at least a decade earlier. Public policy and regulations are largely responsible for this state of affairs. The net result of all this was that the entire framework and the process engendered corruption and wastage at all levels of government

© The Author(s) 2019   275
S. Birudavolu, B. Nag, *Business Innovation and ICT Strategies*,
https://doi.org/10.1007/978-981-13-1675-3_10

and in all walks of life, as is clear from the large list of scams that came to light over the past two decades. Regardless of whether anyone was convicted or not in these scams, the objective evidence for corruption and backwardness are the abysmally low ranks that India has registered progress on several global indices such as ease of doing business, Corruption Perceptions Index, uman Development Index, and so on. The content of these indices is discussed in another chapter, Regional Factors.

There was a massive brain drain of talented youth from India over one generation (last 25 years), as they were deprived of opportunities and employment. The IT sector in India probably flourished because the government never understood it properly and hence did not regulate it to death (Das 2012).

The present government is making an all-out effort to change the narrative and has decided to make sweeping changes to usher India into the new age, improve the lot for its population, and play a major role on the world stage. Many regulations have been passed in the past two years alone that seek to exploit ICT and innovation for pursuing development. This chapter highlights the changes as well as the areas that need to be addressed. While presenting the budget 2018, the finance minister has stressed that the government intends to further the use of latest technologies, such as robotics, AI, digital manufacturing, big data analytics, and Internet of Things (IoT). He then went on to double the allocation on the Digital India Program to INR 3073 crores. India's regulatory regime is evolving and trying to cope with the reality of pervasive technological development. The current chapter briefly highlights the major elements of current policies and policy debates. It also touches India's position on a global regulatory regime which is being negotiated aggressively at the WTO.

USOF (Universal Service Obligatory Fund) was constituted by the government of India in 2002 to raise money for building connectivity in underserved regions in India such as remote rural areas (www.usof.gov.in). Recognizing the fact that industry players would have no commercial interest in laying down network infrastructure in such regions due to the difficulties and the heavy capital outlay required, the government levied an access fee on the Telcos as a percentage of the license fees being paid by them. These fees contribute to the USOF. The government steps in with USOF to provide subsidies and incentivize Telcos to provide services in such regions in India. Currently USOF has about Rs. 48,000 crores ($7.4 billion) available for deployment, having disbursed about Rs. 44,000 crores ($6.7 billion) out of the Rs. 92,000 crores ($14 billion) accumulated by March 31, 2018.

10.2 ROLE OF GOVERNMENT IN INNOVATION POLICY AND REGULATION

Emerging economies have come up now as major incubator for innovation. Many MNCs earn large profit from these economies, and hence more focus is given on localization of products and services and even for developing completely new products. A firm looks at a developing country for possible base for innovation when the following features are satisfied.

A) *Capabilities and Stock of Subject Knowledge*: It should have ability to design products, diversify it, and produce it at much cheaper rate compared to home country.

B) *Capability for Core Products*: It should have the ability to produce the products locally (preferably the entire value chain should be available there but not strictly).

C) *Capability for Modified End Products*: It should have the engineering capability and sufficient component knowledge so that it can modify the end products easily.

D) *Market for End Products*: There exists sufficiently large market for the end products so that risk of innovation goes down due to strong demand of modified products internally.

It is also important to note that national government may also play an important role to encourage the innovation culture in the country. It creates an incentive, subsidy, and regulatory structure which can promote innovation. As innovation by design is mostly a "public good", and there is uncertainty around the innovation process which hinders the pace of innovation, the government wants to play a positive role. There is also a need for certain complementary assets which individual companies cannot possess, and government provides or develops such assets (such as high Internet speed or setting up high-end laboratory at universities, etc.). It also makes certain technology available in the country to accentuate innovation. More precisely, government has an important role in developing innovation environment in the country. The following table explains the economic system and government policy bouquet to promote innovation (Table 10.1).

While developing the regulatory structure, government requires to bring a balance between demand and supply side of the regulation. Demand for regulation may come from public demand to build up the confidence so that companies can invest and take some additional risk. It

Table 10.1   List of agents, activities, and policies to promote innovation culture

Economic systems for innovation (list of agents)	*List of policies and activities by govt. to promote innovation culture*
Supplier	Educator
Manufacturer	Information provider
Complementary innovators	Financer for R&D
Related industries	Provider of complementary assets
Customers	Policy to prevent anticompetitive activity
	Strong patent system
	Holistic regulatory system
	Lead user of technology products

Source: Modified and adapted from Afuah (2003)

Table 10.2   Demand and supply side of regulatory structure of innovation and technology products

Demand of regulation	*Supply of regulation and incentives*
Public confidence on technology	Providing R&D incentives
Pricing and cost consideration by local players	Providing patent protection
Innovation environment development	Fees on complementary assets
Revenue generation for public investment in innovation	Licensing structure on technology products
	Product liability rules
	Maintenance regulation
	Price control
	Differentiated tax Laws
	Product usage policy
	Export incentives

may also be due to create a protection due to cost and pricing concern around the product development process. On the supply side, government brought up regulations by developing incentive structure for R&D, licensing structure for new products, providing patent protection, developing product liability, rule in implementing or aligning new technology products maintenance regulation, price control, differentiated tax laws, product usage policies and export incentives, and so on. The following table provides a snapshot (Table 10.2).

The following sections deal with India's policy environment, and reader can comprehend the logic around such policies based on the above discussion.

10.3 TELECOM POLICY AND REGULATIONS

TRAI (the Telecom Regulatory Authority of India) had issued industry-friendly recommendations for an NTP2018 (New Telecom Policy 2018) in February 2018. These are also intended to increase the Telecom footprint and reduce the digital divide in the country. Based on these recommendations and feedback from the industry and different stakeholders, the DoT (Department of Telecom) has issued a draft policy on May 1, 2018, titled, The National Digital Communications Policy, 2018. The department seems to have used the term "digital communications" instead of Telecom to reflect the broader connotation and transformation in the sector.

The strategic objectives of the policy are:

1. Broadband for the entire population of India.
2. Creation of four million new jobs in the digital communications sector.
3. Increasing the digital communications sector's share towards India's GDP from 6% in 2017 to 8%.
4. Putting India in the top 50 countries relative to ITU's ICT Development Index. Currently India is in the 134th position.
5. Increasing India's participation in global value chains.
6. Protecting India's digital security, because data is a key national resource.

It lays down three missions to fulfill the above objectives by 2022:

- Connect India (to provide universal access/connectivity for India's population)
- Propel India (to improve the industry)
- Secure India (to ensure data protection)

Connect India
1. Ubiquitous broadband at 50 Mbps for all Indians, including fixed access broadband to 50% of the citizens, and 100 Mbps on-demand to all key institutions
2. Unique mobile subscriber density of 55 by 2020 and 65 by 2022
3. All gram panchayats to get 1 Gbps data connectivity through wireless broadband by 2020, scaling up to 10 Gbps by 2022
4. Deploy five million public Wi-Fi hotspots by 2020, and scale up to ten million by 2022

To achieve the Connect India objectives, the DoT is adopting a multi-pronged strategy:

A. *National Broadband Mission to Provide Universal Broadband Access*:
   1. *Funding*: This mission is to be funded and fulfilled through different initiatives such as BharatNet, JanWiFi, and so on, funded by public-private partnerships and USOF.
   2. *Access Network*: Implement a Fibre First Initiative, to provide fiber to homes, organizations, and institutions. Address different aspects such as leveraging existing/legacy assets, promoting collaborative models, providing standards for inbuilding connectivity solutions, and so on.
   3. *Backbone Network*: Establish national digital grid by creating National Fibre Authority, working with public infrastructure authorities, for right of way, project management, and so on, and build open access next-generation networks.
   4. *Mobile Telephony Infrastructure*: Build mobile tower infrastructure, through a comprehensive framework that covers incentives, permissions, exceptions, providing green/solar energy.
   5. *International Networking*: Increase international connectivity through providing international cable landing stations, and promote sharing of infrastructure.
   6. *ICT Convergence*: Promoting ICT convergence through the union of Telecom, IT, and broadcasting sectors in all aspects, regulatory, legal, and technical.
   7. *Metrics for Investments*: Constructing a Broadband Readiness Index for states to attract investments and create right of way.
   8. *Incentives*: Provide financial incentives to encourage investment in digital infrastructure, such as depreciation, tax breaks, and so on.

9. *Innovation*: Encourage innovation and alternate technologies, including virtual network operators.

B. *Deem Spectrum as a Critical National Resource*:
   1. Provision enough spectrum for broadband
   2. Make spectrum allocation a fair and transparent process
   3. Improve spectrum utilization
   4. Encourage next-generation access technologies

C. *Bolster Satellite Communications Technologies*:
   1. Update regulations for satellite communication
   2. Improve utilization
   3. Build ecosystem for satellite communications

D. *Eliminate the Digital Divide by Providing Connectivity in Remote Areas*:
   1. Use USOF to incentivize and fund digital infrastructure development in these areas
   2. Build framework for deploying USOF

E. *Improve Customer Experience*:
   1. Build systems to record and address customer complaints
   2. Prioritize public health and safety standards
   3. Strongly promote use of renewable energy technologies

Propel India

The objective is to improve the industry by enabling next-generation technologies and services through investments, innovations, indigenization, and intellectual property rights generation.

1. *Investment*: Attract investment of US$100 billion by 2022 in the ICT industry.
2. *International Markets*: Increase India's participation in global value chains.
3. *Startups*: Foster technology startups in the ICT sector.
4. *Intellectual Property*: Create IPRs in India that are of international standards.
5. *Patents*: Develop SEPs (standard essential patents) in digital communication technologies.
6. *Skill Base*: Build a skilled manpower base of one million in the new generation skills.
7. *IoT Ecosystem*: To multiply the current IoT base to reach five billion connected devices.

8. *Industry 4.0*: Propel the transformation to Industry 4.0 (which has been explained in another chapter).

The strategies for accomplishing the Propel India mission include:

A. Spurring *Investments* for the digital communications sector
   The strategies range from deeming Telecom infrastructure the status of critical and essential infrastructure to reforming the regulations, simplifying license fees, and facilitating compliance obligations.
B. Fostering a solid approach towards leveraging *Emerging Technologies*
   The plan covers several areas such as charting a path for new technologies such as IoT, AI, robotics, cloud computing, and 5G networks, simplifying the licensing, providing a robust/secure infrastructure, promoting innovation, moving to IPv6, and enabling high-speed Internet, establishing India as a global center for cloud computing and international data centers, enabling smart cities, and exploiting AI and big data in ICT/Telecom.
C. Growing a strong *Research and Development* base
   This strategy has multiple facets such as restructuring the DoT to become a top-notch Telecom R&D center, simplifying processes, and creating frameworks for R&D procurements, licensing, testing, and certification of new products, establishing CoEs, promoting innovation centers and incubators, encouraging Indian IPRs, and obtaining experimental licenses.
D. Encourage *Startups*
   Enable the ecosystem for startups by providing different incentives such as academic and industry collaborations, participation in government procurement, simplifying regulations, reducing entry barriers such as initial cost and compliance burdens.
E. Increase *Domestic Manufacturing*, in quantity, quality, and in value addition
   These strategies include a range of measures such as incentivizing chip manufacturing, easing tax burdens, encouraging indigenization in software and R&D, and compliance to preferential market access stipulations.
F. Build *Capacity* for a strong, large base of human resources
   The idea is build individual and institutional capabilities in all critical areas of ICT, through multiple means. Educational and training resources/material need to be created to facilitate online learning.

G. Bolster *Public Sector Units*

PSUs need upgrading, and their capacity needs to be utilized efficiently in every area, from R&D to production, operations, and training.

H. Catalyze *Industry 4.0*

There is a need to lay out a clear plan to transition to Industry 4.0, by working with industry bodies, create best practices (or borrow them from other countries) build collaborations, and develop markets for the emerging technologies such as IoT and cloud computing for every sector.

Secure India

The goal of this mission is to ensure digital sovereignty and secure all digital communications.

- A comprehensive data protection framework for digital communications that safeguards citizens' privacy, autonomy, and choices and further India's role in the global digital economy.
- Uphold Net Neutrality and implement it across the length and breadth of the structures, including next-generation technologies.
- Put in place robust and flexible data protection frameworks backed by a strong encryption policy.
- Address security issues.
- Build governance and enforce accountability through institutional frameworks.

The policy aims to accomplish these goals using a variety of means such as developing security standards, policy on encryption and data retention, aligning the legal and regulatory frameworks as relevant to the security standards, being part of global standards bodies, setting up a security incident management and response system, building a sectoral CERT, creating and enforcing SOPs, developing a pan-India disaster relief program and allocating spectrum and networks for it, and fostering digital infrastructure sharing mechanisms.

Overall, the draft policy also takes note of the severe crisis in the Telecom companies, and the deteriorating health of the debt-ridden sector. The industry forums have long requested for a friendly policy that is beneficial to industry, investment, and innovation, and bring stability to the sector. With policy emphasis on Net Neutrality, the Telcos need a more level playing field with the OTT players, to be able to get a decent ROI on their investment and enable them to upgrade their networks. This new policy should help India align with the global best practices, to help the Telecom sector transform India.

The Telecom industry was disappointed that their key recommendations were not taken into account in the Finance Minister's 2018 Budget. Currently the Telecom companies pay to the government levies totaling to about a third of what they earn.

The recommendations that the industry had urged the new policy and the budget to include urgently are (Telecomera 2018; Parbat 2018):

- Reduce GST on Telecom services from 18% to a maximum of 12%.
- Cut import duties in half on finished mobile phones and imported components from the proposed 20% and 15%, respectively, because the overall manufacturing costs are only going up, to the detriment of the Make in India program and the interests of the population.
- Elimination of basic customs duty on 4G LTE equipment, to reduce network rollout costs.
- Clarity on right of way tax at the state level.
- Need to lower tax to 1% on discounts extended to small dealers.
- Telecom finds no mention in the budget under infrastructure, unlike roads, railways, electricity, and so on. Telecom, being the backbone of the digital highways and digital economy, should be treated as infrastructure.
- Need to cut Telecom levies for SUC (Spectrum Usage Charges), license fees, and USOF (Universal Service Obligation Fees)

Most of these can now be implemented using the framework in the new draft policy.

The EU, the USA, and Japan have been claiming that voice over protocol phones and optical transport equipment are covered under the ITA-I and should be made duty-free by India. However, India is opposing them on the grounds that these products did not exist when the agreement was signed and, hence, are outside its coverage though the items might be falling under the codes that are covered under the ITA-I. For example, the definition of Telecom equipment has itself changed as it now includes new range of technology products. Under ITA-I, the code for Telecom equipment (HS Code 8517) referred to telecommunication apparatus for "carrier-current line systems and digital line systems". In these last few years, various new technologies have been developed. Subsequently, the code has been expanded to include these new developments. However, ITA-I obligations cover only those items that confirm to the then definition of products for "carrier current line systems and digital line systems". India gave its views on this and stated it cannot consider these developments as part of ITA-I. Also, India has decided not to sign ITA-II after it had a bad experience with ITA-I. It adversely affected India's electronics and IT hardware industry which could not realize its potential due to cheaper imports coming in from countries such as China, the USA, and South Korea. It almost wiped out the IT industry from India especially the hardware. It is also important to note that India is now embarked on having a sound domestic hardware industry through "Make in India" and mushrooming of tech-based SMEs is shaping its ecosystem. A robust innovation policy and congenial regulatory system is a necessary for the domestic growth of electronics and ICT industry. At a high level, these also seem to be taken into account in the new policy, and it is hoped that the actual implementation will lead to an extremely vibrant ICT ecosystem in India.

10.4 Open-Source Software and Open Standards

MeitY (Ministry of Electronics and Information Technology) has a policy on adoption of OSS (open-source software). It stipulates that OSS be considered first prior to any other commercial software. If the latter is chosen, it will need a justification as to why OSS did not suit the purpose. With the advent of standard/open architecture frameworks becoming the norm in every area, interoperability based on open standards became a priority in the industry. The open standards in all the areas ensured quality, interoperability, compliance, ease of upgrade, and so on. Most enterprise software products today cater to open standards; otherwise they can't even bid for deals, because open standards are an entry criterion. Thus even commercial software that adopts open standards and has a good record should be given preference on par with OSS. This is a good move for India and the Indian industry as a whole, because the basic OSS is getting better and stronger by the year, and has reached industrial grade in terms of functionality, usability, robustness, security, scalability, availability, and performance. Examples of basic OSS are operating systems (Linux, Android), HTTP servers (Apache), development tools (Python, Eclipse, mySQL), containerized applications management systems (Kubernetes), web browsers (Firefox), distributed computing (Hadoop). It comes at low cost, avoids vendor lock-in, has good interoperability provided by the industry, follows standards, and is improved upon by thousands of developers, organizations, and institutes worldwide.

Most of cloud computing runs on OSS. Indian industry will greatly benefit from OSS because there is a much wider pool of talent to use from. For example, if an engineer is trained in Linux, he can be useful in any sector and in a wide variety of jobs. Thus services industry profits immensely: the OSS has rock-bottom costs, talent is widely available, and demand is high. For an Indian product development company, it gets the entire software infrastructure and tools, for free to build its product. Upgrades and bugfixes are also released worldwide, so there is adequate assurance for the future that the software will be supported strongly.

10.5   ARTIFICIAL INTELLIGENCE

The government of India has set up an expert panel in October 2018 to advise the ministry on AI (Artificial Intelligence) policy. The government has also drawn up a seven-point strategy for AI, which includes man-machine AI interactions, creation of an AI-savvy workforce for AI usage and research, ethical/social/legal aspects of using AI, measuring AI (establishing benchmarks, standards, practices, etc.). Several other countries (USA, UK, Germany, etc.) are also developing policy on AI.

One key goal of this proposed policy is to prevent AI-powered cyber attacks. AI has the potential to add US$957 billion to India's economy (Accenture 2017). The primary reasons, apart from improving cybersecurity, for proposing an AI policy are fillip to startups, job creation, Make in India program, and improving quality of life.

This is a welcome move for the Indian industry which would not like to be left behind on the global stage, because the entire world is now keenly looking to adopt AI for all the reasons explored in the chapter on technology. This will however imply training/re-training of millions of engineers and establishment of new practices, such as data/information sciences, consulting, training of AI systems, and so on. For Indian companies become leaders in the digital age, there could be no better opportunity than this.

10.6 IoT (INTERNET OF THINGS)

India is on an aggressive path to modernization, and IoT's adoption will explode in the Indian industry, for governments, enterprises, and consumers. It is a market worth trillions of dollar opportunity that leverages electronics, software, and communication, and is driven by four factors:

- Fall in the price of sensors, computing resources, and communication
- The dire need for automation of machine operations and logistics, to save labor and capital, cut wastage, improve utilization, and so on
- Rise of new technologies, industry standards, and frameworks
- Exciting growth in new business models that seamlessly bring together people, machines, and resources to create new value streams

The IoT technology itself is discussed briefly in another chapter. The Indian government has declared that it will be developing 100+ smart cities. This will need IoT on a massive scale. Some of the applications will be water management, energy management, waste management, smart lighting, transportation and logistics, law and order (public safety), and so on.

To be prepared for this scale of change, the government has embarked on creating a policy for IoT (MeitY IoT 2016). This policy draft document raises the following topics:

Open platforms for IoT, infrastructure capacity to handle IoT traffic, research and development, standards and frameworks, education and training on IoT, quality and costs, advisory committees and policy making bodies.

The draft policy identifies the five pillars for IoT (the first five listed below), on top of which governance structures and standards are built.

1. *Engagements and Initiatives*: To attract venture funds into IoT technologies such as sensors, memory, processors, solar electronics, and so on, and to help promote export of IoT products and services. Events, fairs, and programs will be planned, as also participation in global forums.
2. *R&D and Innovation*: To fund common R&D that will be beneficial to all, in terms of access to a solid base of technology, collaboration, readily available cloud-based open-source projects, testing labs, and so on. This is also to stimulate investment from private sector into IoT projects in India.

3. *Capacity Building and Incubation*: To build India's capacity in IoT at an institutional level. With ERNET as the nodal agency, 15 academic/institutional partners will develop IoT centers, where the community can experiment with IoT devices and applications.
4. *Demonstration Centers*: To develop domain-specific and high-priority strategies and prototypes, such as smart grid, logistics, automotive, and so on. This is also with a view to fund them in a PPP (public-private partnership).
5. *Human Resource Development*: To create awareness and talent in IoT, to develop skillsets and knowledge useful for the industry and academia. For this the government plans to facilitate academic programs right from basic training to master's/PhD level, and will create bodies in academic, industry, and associations for setting up cross-connects, labs, and centers of excellence.
6. *Standards*: To promote IoT technologies and expertise in the country and develop standards and quality. This relates to technology, process, interoperability, services, for example, standards pertaining to communication protocol, architecture, energy, safety, privacy, and so on.
7. *Governance Structure*: The draft policy takes a four-pronged approach to governance, which includes legal framework, advisory committee, governance committees, and a program management unit.

10.7   DIRECT CARRIER BILLING

With mobility (and smartphones) increasing in India, and its reach far exceeding that of landlines or of PCs, it is imperative to leverage it well for masses. With demonetization and fall in mobile data pricing (accelerated by launch of Reliance Jio), mobile payments should be made all the more attractive, as it has a direct impact on the economy.

Direct carrier billing (DCB) is an online payment method in which the *Telecom subscribers* can pay for services or goods purchased, by adding those charges to their phone bill. India's DoT (Department of Telecom) allowed DCB for up to Rs. 20,000 for both post-paid and pre-paid subscribers. Adding DCB for pre-paid subscribers, from 2017, was a significant move as these constitute 95% of India's mobile subscriber base.

The current regulations however need changes because there is a disincentive for the operators to deliver DCB. The Watal Committee Report (Watal 2016) clearly states that "The present legislation does not permit DCB (direct carrier billing) payment model".

The controversial part is that the DoT stipulated that revenues from DCB will not be considered as pass-through revenues for computing AGR (Annual Gross Revenue) for license fee and spectrum fee.

This is a major blow to the mobile Telecom operators for supporting DCB, despite their willingness (or self-imposed compulsion) to continue to offer DCB. This is because the government of India taxes the Telecom operator's revenue for license fees and spectrum usage, both together bundled as WPC charges (named after the Wireless Planning and Coordination Wing of the Ministry of Communications), which typically constitute some 10–12% of the Telecom operator's AGR.

In the case of DCB, the Telecom operator and the content aggregator keep their cuts, that is, share of the revenue before passing on the money to the content provider (or the goods/services provider as the case may be). Thus the higher the AGR, the greater the amount of money lost at every stage before the original content/goods/services provider gets his due. This is a great disincentive to the entire DCB value proposition.

As of this writing, it is hoped that the regulations (NTP 2018) will change for the better to massively incentivize direct carrier billing, which will allow businesses and services of every kind to tap into the bulk of India's population with relative ease. To start with, they can treat the DCB as "pass-through revenue" for calculating AGR. This will energize the entire value chain, and stimulate innovation and business in every part of the short head and the long tail. The current regulations for DCB favor post-paid mobile subscribers rather than the pre-paid ones.

UPI (Unified Payment Interface) will help India move towards a cashless economy by enabling money transfer more easily from mobile phone, without the need for a mobile wallet. But DCB is also needed, because an entire system is already set up by the mobile operator. And vendor collaborations with the Telco operator are easier. There is also lesser risk of fraud, because the vendors can be validated by the Telco operator.

10.8   Mixing IP-Based VoIP Calls with PSTN/ Mobile

In India it is forbidden to mix VoIP on Internet calls with PSTN/mobile calls. With improvement in both quality and access to the Internet, voice over IP is gaining traction. Using applications like Skype, WhatsApp, Yahoo Chat, Cisco Webex to name a few, one can make good-quality voice, video, and even conference calls over the Internet. However bulk of the Indian population relies on mobile networks and PSTN (public switched telephone network), that is, the traditional landline telephone networks. In India it is not permitted to mix VoIP and PSTN calls, that is, one cannot make a phone call from the Internet to a PSTN/mobile number or vice versa. In technical jargon, the call initiation and termination need to be on the same side, either on PSTN/mobile or on the Internet.

VoIP on the Internet is considered the same whether originating from (or terminating on) mobile device, laptop/desktop computer, or any other device. It should be noted that the restrictions in India apply to voice calls only and not to text or multimedia messaging. One may have a website to send bulk SMS messages to mobile subscribers. Obviously, in this case, the messages originate on the Internet and terminate on the mobile number.

Most technologically developed countries do not have *any such* restrictions at all; hence a lot many (and richer) technical and business models are possible.

Whereas in India, the restrictions on mixing ostensibly seem to be there to protect the Telcos' revenues from being cannibalized by the OTP (Over-the-Top) VoIP players. However this artificial restriction has become quite burdensome, especially when IP pipes are very efficient nowadays. Telcos have in fact resorted to very roundabout means to meet the regulations while leveraging the IP networks, like using an IP network for transporting the IP packets for bulk of the route.

The interconnections on either side are typically achieved through SIP Trunking. This means the call travels on IP network for *most* of the route, and only the end-points are shown to be on mobile number/PSTN just for regulatory compliance! Sometimes the architecture is even more absurd as given below. If one makes a call to a foreign country and the interconnection is done by a box on foreign soil, where Indian regulations do not apply, then the terminating point can be on IP network in India.

This means that the interconnection box is kept in a foreign country, where Indian regulations do not apply! These evasive methods should not be necessary at all.

There are many important reasons why regulations should be changed to permit mixing of VoIP calls and PSTN/mobile calls, even if it results in a short-term dip in PSTN/mobile revenues:

1. *Efficiency*—IP networking and SIP Trunking work on a connection-less protocol, hence are far more efficient. This means that unlike connection-oriented phone calls, they don't lock up bandwidth for the duration of the call. In VoIP, the call is chopped into IP packets and re-assembled in real time at the other end. Hence the full bandwidth of the IP pipe is available at all times for transporting the packets. Thus, one can pack the pipe to the full capacity. Internetworking should be allowed in a straightforward manner.

2. *Rich Services*—The Internet offers tremendous flexibility and power. The level of automation, granularity, and workflow obtainable with IP networks is unimaginable with plain PSTN/mobile networks, primarily because the former is all built on software. Hence much richer services are possible through innovation and collaboration. This will offer a superior customer experience. With AI and Machine Learning, a lot more sophistication can be brought to the customer experience.

3. *Scale*—The Internet can offer services that are on the scale of the planet! The normal mobile and PSTN networks cannot do this. The scaling is possible because of the fat, gigantic backbone networks and huge data centers. Software can be distributed and scaled up in many ways, right from improving the algorithms to adding servers.

   For all the reasons mentioned above, the regulations should allow mixing of IP and voice calls. The Telecom industry may take a temporary hit, but can come up with far richer, efficient, and scalable offerings.

10.9   Cybersecurity

As discussed elsewhere in this book, cybersecurity is paramount in the Information Age. MeitY has released a white paper on data protection framework for India on November 28, 2017. This whitepaper is intended to pave the way for a legal framework for data protection and has the stated goal to "ensure the growth of the digital economy while keeping personal data of the citizens secure and protected". The government recognizes that data protection is the foundation on which data-driven innovation and entrepreneurship can flourish in India. The twin themes of the paper are:

- Secure digital transactions, and
- Data privacy

The broad topics it raises are those related to consent of consumers while collecting their data, disputes and resolution, penalties and compensation, enforcement model, and so on. The sensitivity of information may increase with the number of pieces of data collected. For example, knowing the full name of a person may be harmless, but as one keeps accumulating other pieces of information, such as date of birth, postal address, email address, and so on, it can badly compromise both identity and access to his data everywhere. It gets easier to progressively gain access to all of the person's private data with piece by piece incremental access to information. Big data analytics and data mining aggravate the situation, because they can uncover patterns from seemingly unrelated bits of data, for example, location of a mobile device at a particular time, and credit card transactions record can put together an entire travel history, spending patterns, and habits of a person and his family members. Selling these data insights to different vendors because of commercial value can be a serious breach of privacy.

The paper intends to create a legal framework with guidelines for all the different actors to operate in, and put measures in place for monitoring, reporting, and redressal. In special cases, the government needs access to data for reasons like law and order enforcement and national security.

The government of India had launched a Botnet Cleaning and Malware Analysis Centre, because of which the infections have reduced by 51%, as per government reports (Ghoshal 2018). The malware in banking sector reportedly fell by 74%.

Cybersecurity laws and legal frameworks are a boon for the Indian industry because cybercrime will cost businesses globally over US$2 trillion by 2019 (Juniper Research 2018). As India seeks to improve its cross-border digital trade, it needs assurance of security and stability.

10.10   CERTIFICATION OF SOFTWARE AND HARDWARE EQUIPMENT

With rapid advances in technology and an explosion in a number of software/hardware offerings in a competitive market, it has become extremely challenging to select products. Enterprise product purchase decisions are enormously complex and have far-reaching implications. Setting standards and verifying quality is of paramount importance for a country, because mistakes imply an organizational and national wastage of resources. Bad purchase decisions can adversely affect the economy, security and well-being of a nation. Organizations are confused due to the severely competitive sales pitches and lobbying by the vendors from different countries.

Consider enterprise ICT products such as Telecom switches, optical network equipment, routers, session border controllers, and high-end servers. Any purchase decision affects the lives of millions of that country. A Telecom switch bought from foreign corporation could compromise national security either intentionally (systematic spying by enemy country) or unintentionally (due to poor design), by tapping into or leaking information pertaining to the private/official communication among citizens and officials holding key positions in the country. When a Telecom switch is substandard, it could again cause widespread damage and wastage due to down-time or by choking up on the communication among thousands of people, for example, through dropped calls, blurred voice signals, or plain inability to connect, especially during peak hours.

The bottom line is that ICT systems (hardware and software) have a huge multiplier effect on business and society, and hence their specifications, implementation, and quality cannot be taken lightly.

The government has stepped in to set up regulations, standards, and certifications.

Conformity assessment lays down that certain hardware equipment need to have certifications to be compliant, for example, EMC (electromagnetic compliance). India is part of the CC (common criteria) authoring nation status for ICT. Thus, India's testing labs will match global standards, and so they can do quality checks and issue their own certifications.

Conformity assessment requires that the manufacturer of the listed ICT product (not designer) register the product with the BIS (Bureau of Indian Standards), provide samples to BIS laboratory for testing, and provide an undertaking, that is, a self-declaration that the product meets BIS standards laid down for a list of particular ICT products.

The conformity assessment regulations will cover rules for repair hubs, testing, spares/repair process, technology transfer, encryption standards, and re-import of used spares,

The government takes recommendations from industry bodies such as COAI, MAIT, NASSCOM, and so on. It has set up quality control laboratories and standards bodies in organizations such as C-DAC (Center for Development of Advanced Computing), NIC (National Informatics Center), DRDO (Defence Research and Development Organization), CSIR (Council of Scientific and Industrial Research), and NCTF (National Cyber Task Force).

The security requirements for Telecom equipment and some other ICT products are quite stringent, and they are likely to get even tighter. There is mandatory security testing for all Telecom/ICT products. The global standard bodies and OEM (original equipment manufacturer) will be accountable for implementing all these policies.

With the advent of such regulations, not only will the quality, efficiency, and standards in the industry improve, but a host of opportunities will open up for licensed vendors to provide training, certifications for both people and equipment, and consulting and testing. Hence it will have a salutary effect on the Indian industry.

10.11   CRYPTOCURRENCIES

Bitcoin (BTC) is not recognized as legal tender in India. BTC may be obtained in India either through mining, or by direct purchase on the net in exchange for real (legal) currency, or as consideration for goods and services sold. The Indian government needs to squarely address the topic of cryptocurrencies (including bitcoin), and is still working on developing the relevant regulations. Despite the advantages, there are several concerns with cryptocurrencies like bitcoin:

- *Anonymity*: Governments and banks require people to be clearly identified in transactions, for many reasons like taxation, law and order, collecting economic data for making policies, and so on.
- *Illicit uses*: A large portion of bitcoins are being used in gray areas, questionable cross-border dealings, murky/illegal transactions including piracy, arms, drugs, malware, and every kind of crime

imaginable, without fear of any authority or of traceability, and with full guarantee of verification of transaction.

- *Inefficiency:* Wastage of energy due to the mining and peer-to-peer process, as compared to the latest non-bitcoin-based financial transactions such as electronic wallets, payment gateways, online banking, electronic funds transfer, credit cards, and debit cards.
- *Lack of control:* This is a double-edged sword. There is no governing authority, but an algorithm, and there are no borders. The pertinent problems have been discussed elsewhere in the book.
- *Security:* Problems related to hacking, theft, and hard forks have not been adequately addressed globally. Legal disputes are hard to resolve.

However, all these have not prevented widespread acceptance of bitcoins, predominantly for speculation and other transactions. Bitcoin exchanges have sprung up and there is a lot of representation from India. A clear-cut policy needs to evolve in India, or there will be chaos and problems. It should be noted that even several of the existing, thoroughly regulated systems are used for speculation and are amenable to rigging, for example, the share market, commodity trading, foreign exchange trading ($5 trillion traded *daily* globally, 90% of which is speculation).

In December 2017, the value of 1 BTC rose to over US$19,000 and hovered around $15,000. With such high valuation for a cryptocurrency whose fundamentals are difficult to fathom (because they emanate purely from algorithms and are unregulated by governments), it attracted the attention of the governments worldwide. The income tax authority of India has sent notices to HNI (high net-worth individuals) who are holding BTCs. It is likely that BTC will be brought under the ambit of the law and will be treated as wealth, with applicable wealth tax and the profits from BTC transactions treated as income, subject to income tax. Several other countries like Japan, the USA, the UK, Germany, Australia, Brazil, Canada have already taken such steps.

So far there are very few legitimate business models in the industry that rely on bitcoin. Hence the stringent laws pertaining to bitcoin are in order. The industry should not be affected adversely by these laws.

10.12   BLOCKCHAIN

The blockchain technology, which bitcoin is built on, is being re-purposed for several uses. Some of these use-cases that may suit India well are:

- Land and real-estate records—these are often manipulated and disputed, especially where the stakes are high. Real estate in India has become a haven for black money. The transactions need to be certified and be locked in immutable ledgers. Blockchain may fit the bill.
- Quality certifications and tests of the most key components in critical industries like aerospace, pharma, nuclear power, and so on.
- Government-issued tenders and contracts that should be open for RTI (right to information) inspection.
- Very high-valued financial transactions for assets like shares, stocks, bonds.
- Some critical legal cases and legal matters, where transparency is needed.

These indicate areas that are susceptible for manipulation, fraud, and corruption. In the global Corruption Perceptions Index, India ranks 79th out of 176 countries, clearly indicating the need for much change and clean-up. Used properly, blockchain can help provide immutability, transparency, and verifiability of transactions.

Blockchains need not have anything to do with cryptocurrency, although the currently popular public blockchains need cryptocurrency to incentivize mining/validation. Thus, the non-cryptocurrency blockchain implementations do not need really need regulatory approval to get started on, if there is adequate agreement among the stakeholders. They may be treated as plain ICT projects to begin with. However, it will be good to fit in regulatory compliance right from the start. Many of the projects are, in fact, in the government sector as is obvious from the above examples.

Some areas may still need regulatory guidelines: legal agreement among stakeholders (especially in smart contracts), how data is dealt with, how it is amenable for audit, and whether it needs to reside within the geographic boundaries of India, legal jurisdictions, and guidelines in case of problems and disputes.

The Indian industry is embracing blockchain in many sectors. The use-cases for blockchain are discussed in the chapter on technology. Government regulations will help, for example, in matters related to legality of smart contracts, ownership of and rules for updating of blockchain application software and business rules such as putting in bugfixes and new features, and so on.

10.13   DRONES

Drones are UAVs (unmanned aerial vehicles) or RPAS (Remotely Piloted Aircraft Systems). Due to safety risks, the Indian government banned drones for public use. However, on October 30, 2017, India's aviation regulatory body, DGCA (Directorate General of Civil Aviation) released draft regulations for drones. In India, drones can be very useful for a variety of uses, such as:

- Timely transport of critical medicines and blood especially to remote/unreachable areas and even the crowded areas that are hard to reach quickly.
- Survey of disaster affected areas, and dropping off relief supplies.
- Combat poaching, intrusion, trafficking, terrorism, and so on, especially when used in conjunction with other systems such as automatic surveillance/monitoring systems.
- Crowd control by police.
- Photography, advertisement film making.
- Oil and gas exploration and for improving agriculture.
- E-commerce deliveries of packages. Already the e-commerce companies have petitioned to the government of India. Restaurants can also use it for food delivery to homes.

The newly released draft for drone regulations has classified drones into the following five categories based on the weight of the drone—Nano, Micro, Mini, Small, and Large. These range from less than 250 grams (Nano) to greater than 150 kg (Large). Barring the Nano, all the others need to be registered with DGCA and get a UIN. Above Micro, they also need an air defense clearance.

Remote pilots for all drones need to be adults who are registered and trained as specified by the DGCA process.

The flight route of any drone cannot be changed, once it has been approved. Furthermore, drones are prohibited from flying within 5 km radius of specified areas, such as airports and parliament/Rashtrapati Bhawan, and within 50 km of international border or LOC. All densely populated areas and eco-sensitive areas like national parks are also out of bounds for drones. All drones need to adhere to privacy protection laws and they cannot film anyone without permission.

These regulations will need to change as smaller and smarter drones keep evolving, and can be used for military purposes, such as lethal autonomous weapon system (LAWS) (Russell). The proposed regulations are timely and welcome. The industry can look forward to developing many use-cases with drones. Some shortcomings of the regulations (TRA 2018) from the industry's point of view are:

- Issuance of UIN and other clearances could become a lengthy and painful process.
- The line-of-sight rule for drones can become quite restrictive for several uses, such as videography and journalism,
- Guarding privacy is a challenge and more regulations are needed.
- Drones needn't have the full extent of laws that aircraft are subjected to, because drones operate at lower altitude. However there are other areas that drones can infringe on private property, spy on others, and so on. These also need laws to abide by, or to settle disputes.
- Ownership of airspace needs to be defined more clearly; its currently very fuzzy.

10.14   NET NEUTRALITY

India's Telecom regulator, TRAI, released its recommendations supporting Net Neutrality on November 28, 2017. It reiterated its earlier stand of February 2016, in which it rejected Facebook's Free Basics, a program intended to be launched in India, but which went against the grain of Net Neutrality. Facebook did this by seeking to provide to the public free access to the web *but restricted in scope* (in terms of number of websites). TRAI's intention is to ensure that all the customers in India get "unhindered and non-discriminatory access to the internet", thereby following the guidelines laid down by the United Nations in 2011, which essentially declared access to the Internet to be a basic human right, and not a luxury. TRAI's regulation will mandate that Indian Telecom operators cannot control either the internet content or the speed at which the content traffic flows. Some exceptions have been allowed such as providing emergency services, dealing with security issues, or facing unpredictable traffic congestions.

With Net Neutrality firmly in place, India can look forward to the Internet remaining a level playing field and a super platform for communication, innovation, business, governments regardless of whether one is an individual, a startup venture, large corporation, a government, a social group, or even a machine (IoT/AI/ML/Drone/Robots/OTT). There will be no discrimination among the different data packets that flow on the basis of content, origin, destination, or use. It is a great step forward for India's digital economy and will encourage the development of apps of all kinds in every area, and boost the development of ICT, business, innovation, and customization. Understandably, Indian IT industry's apex body, NASSCOM, has hailed TRAI's ruling.

TRAI's decision is a wise one because India is a large, diverse country and a developing economy with millions of small, emerging players for whom the Internet offers a spectrum of possibilities, right from finding/ matching jobs for individuals to selling online. If a platform like the Internet itself goes commercial, then the entire nation will be held hostage by the large players, many of whom could be foreign corporations (or even foreign governments) with deep pockets. It will absolutely choke the creativity and growth of all the small Indian players, whose traffic may find it difficult to reach their desired subscriber/consumer base, causing many to give up their efforts over time. A few heroic successful small ones may have to remain subservient to the large players and get taken over or fall in line with the latter in many ways. The real danger is that the Internet is so great in its reach and power that if a handful of monopolies control it, they can sway a nation's opinions and life of millions in unimaginable ways. This will be detrimental to the interests of India from the viewpoint of development, security, growth, and sustainability.

Monitoring the networks for violations of Net Neutrality is a perpetual challenge because there are many layers. There need to be sophisticated instruments and metrics to collate data from the trunks, access networks, pricing plans of the provider and about the speed of data, both from end-to-end and at every segment. Big data analytics can put together a comprehensive picture about compliance to Net Neutrality.

TRAI also needs to include in its regulatory purview, all the players responsible for the network:

- ISPs (Internet Service Providers),
- Content Providers, and the
- CDNs (Content Delivery Networks), who sit between the above two

Going by the regulations for Net Neutrality, TRAI seems fixated only on the first one. It needs to expand the scope to cover the other two as well.

Blatant violation of Net Neutrality is easy to catch if the operators openly publish/market differential pricing for different types of traffic. But violations can also be done in subtle ways by any of the three players, by selectively prioritizing one type of traffic over the other. TRAI needs to monitor and enforce all types of possible violations, because Net Neutrality is so important.

To do this, TRAI should take steps on:

1. *Technical Expertise*: Form groups of technical experts to determine how to monitor networks and lay down compliance standards.
2. *Monitoring Tools*: Encourage the ecosystem to develop tools that monitor traffic and violations.
3. *Governance*: Set up independent monitoring and governance centers that analyze traffic.
4. *Compliance*: Demand that different network components (both hardware and software) adhere to the standards, and enable auditing, monitoring, and governance.
5. *Talent Pool*: Increase expertise and talent pool in the market for technical skills and governance for the above standards and tools and to man the governance centers.
6. *Complaints Resolution*: Enable easy mechanisms for reporting, fast complaints resolution, and audits/penalization. For example, if a web vendor feels that his traffic is being discriminated against, he should be able to file a complaint, and the relevant people should be able to pull traffic reports easily and analyze them, report violations, and send legal notice, penalize, and ensure redressal.

With Net Neutrality in place, the government has brought in stunning legal clarity in one bold stroke, instead of saddling everyone with perpetual guesswork, and companies getting dragged into many court cases. The courts and the government also need not deal with problems on a case-to-case basis. The government must also think of measures for creating a level playing ground so that the Telcos can compete with the OTT (Over-the-Top) players. For the Telcos, although they will face the burden of having to uphold Net Neutrality, the flip side is that Net Neutrality will make it possible for a multitude and variety of long tail type of business models to flourish. This is quite beneficial not only to the Telcos but to the entire ecosystem. Actually, most Indian Telcos have supported TRAI's decision on Net Neutrality while merely protesting on the modalities of implementation (Kak 2017).

10.15 STORAGE OF DATA WITHIN INDIAN BOUNDARIES

In April 2017, MeitY released guidelines on IT Infrastructure for government departments. It stipulated that all data related to government (central/state/district/municipality levels) *must reside within the geographical boundaries of India*. As government may rely on contractors who are cloud service providers to implement their cloud solution, the guidelines state that:

> The terms and conditions of the Empanelment of the Cloud Service Provider has taken care of this requirement by stating that all services including data will be guaranteed to reside in India.

Only the protection of SPD (sensitive personal data) carries legal liability and not any other PD (personal data). SPD includes biometric data, passwords, and personal medical records. Collection of SPD needs consent from the user. Other data such as photos (barring any biometric photos), personal contact lists, phone numbers, credit histories, and so on do not constitute SPD and hence are not covered by these protection rules.

While the move may be good when applied to government's data, in view of sensitive areas such as defense or citizens' data, the industry has several concerns on this stringent act (Regidi 2017):

1. Huge cost implications to establish local servers and the software setup.
2. No longer have easy access to advanced technologies and services, because those are present on the servers on a foreign soil.
3. Less access to global services for consumers.
4. No guarantee of protection because even localized servers can be compromised. In fact, some well-protected data centers abroad may be safer than locally protected servers. Protection implies both physical and cybersecurity.
5. Data processing involves multiple roles: the controller, who collects data for the first time, and processor, who handles and processes

data subsequently. The regulation puts the entire onus of data security on the first party, the controller, and not the processor. For example, if a Telecom operator collects the details of its subscribers, it is the controller. If it outsources the processing of the data to a systems integrator, then the latter is the processor. In India, as per the new regulation, the processor has no legal liability in case of a data breach, but the controller does, whereas in Europe both are jointly and severally liable. This puts liabilities on an unequal footing, which will have to be carefully bridged by the contract signed between the controller and processor.

10.16   Cloud Computing Policy

In June 2016, TRAI released a consultation paper on cloud computing in India, because cloud has become all too important in ICT.

The cloud computing service enables a pay-as-you-go model, in which organizations can use computing resources, such as software, processing power, storage on a need basis as per their requirements, instead of investing in setting up an entire IT infrastructure for their use. The latter would need a heavy outlay of capital, lock the user into specific technology, and would impose the burden of operational costs and maintenance on him. As the systems age, the maintenance would become harder owing to outdated components and old versions of software. Cloud due to its flexible nature can be used to launch and provision new e-governance initiatives. The cloud computing guidelines fall in line with the cloud computing policy called MeghRaj policy, to provide cloud service to government departments (GI Cloud).

The TRAI recommendations to the DoT (Department of Telecommunications) include all aspects, including legal and regulatory framework for cloud computing services in India, cloud platforms, interoperability between the platforms, security of data and operations, government initiatives to promote cloud computing, and so on.

The key areas are:

Data Ownership

The author of the data retains ownership rights, but the Telcos may work out co-sharing agreements. The CSPs are usually quite powerful, so they may practically coerce the ordinary users to sign a ULA (user license agreement) to have them agree to share the data. Unless the users or enterprises are very legally and technically savvy, they mostly end up signing the agreements in favor of the CSPs. And the CSPs in turn use the data for mining or selling it to advertisers and other companies. TRAI's recommendations include data ownership rights and that CSPs should offer options should the user decide to move his data to another cloud service provider.

Data Privacy and Data Protection

The cloud service provider is required by the Information Technology Act, 2000, and the Internet Intermediary Liability Rules (2011) to protect data privacy for the personal information of its cloud users. The TRAI paper stipulates that the users be kept informed of all the risks that they're exposed to, as well as about the protection measures taken by the CSP to safeguard the user data. This should include operations that may be done by the CSP to manage the data.

Cross-Border and Multi-jurisdiction Issues

Governments across the world are concerned about sensitive information going outside their geographic boundaries, for various reasons:

- Other foreign governments, on whose soil the data resides, could access that data and use it in unknown ways to the detriment of this nation, for example, sensitive information related to defense, taxes, intelligence and law and order related, and government contracts.
- The legal jurisdiction shifts to another country, and the other governments may have different laws pertaining to security or privacy of data.

Cloud computing facilitates all-pervasive, global, convenient, 24x7 on-demand access over the network to a pool of computing resources. This in turn means that data may be stored, replicated, or moved around in ways that make it easier to facilitate cloud computing. And this may happen without the knowledge or approval of the end-user. Hence it complicates matters pertaining to security, privacy, and cross-border jurisdictions. The problem is also compounded by the fact that cloud providers may outsource the responsibility of managing data to other vendors who run data centers. There are multiple layers and intermediaries who transmit, handle, and store data.

The Indian Act, Reasonable Security Practices and Rules, 2011, puts the onus of protecting privacy of personal data on the organization that is responsible for handling the data, and lays down penalties for breach/

fraud/tampering of source code and stipulates that user consent be taken for the way the data might be used or disclosed to a third party. As discussed in a previous section, the Indian government also intends to propose a broader right to privacy bill that will empower individuals' rights to privacy of their data and ensure that critical data remains within the borders of India.

TRAI's paper on cloud computing takes all these into account and actively refers to them.

However, this paper should be considered the first step, and the long-term vision should include the principles and provisions of other such evolving frameworks across the globe, such as the USA's Privacy Shield and GDPR (General Data Protection Regulation) from the European Union.

Access by Law Enforcement Agencies

The Indian government may seek access to user information stored on the cloud or even tap information in transit for law enforcement purposes. In the case of encrypted data, the law enforcement agencies may seek decryption keys or even ask the CSP to build a backdoor access to the cloud data. The CSP is required to comply.

There could be difficulties in the case of foreign cloud service providers as they fall under a different jurisdiction and may not cater to the requirements of the Indian government. The government may need to get blanket formal agreements from the overseas CSPs to get access to the data of interest.

Cloud also bolsters the RTI (Right to Information) Act because data can be retrieved instantly against queries.

10.17   Universal Accessibility

The Universal Electronics Accessibility policy passed in 2013 (DNIS 2013) is aimed at improving the lives of the disabled by providing equal unencumbered accessibility to ICT products, both hardware and software. The policy also recognizes the need for regionalization, that is, local language support for the ICT/electronics products. It encompasses both technical and non-technical services, and a universal, friendly design that the disabled can access easily.

10.18 ICT AND INTERNATIONAL REGULATORY ISSUES: INDIA'S POSITION

10.18.1 Information Technology Agreement (ITA)

International trade of ICT products is mostly governed by Information Technology Agreement (ITA). A plurilateral agreement at the WTO Ministerial Conference in Singapore in December 1996 was signed by 29 participants [including the QUAD countries (the USA, Canada, Japan, and the EU), Singapore, and Hong Kong] which paved the way for ITA-I. The Agreement came into force in 1997. The number of participants has now grown to 82, which accounts for about 97% of the world trade in IT products.

As most of the manufacturing process and services are now significantly dependent on IT products for automation, digitalization, payment, delivery, and so on, ITA was largely accepted by the world community. ITA-I focuses on the expansion of world trade in information technology products realizing its importance in the development of information-based industries and in the dynamic expansion of the world economy, raising the standards of living and expanding the production of and trade in goods. The agreement aims to bring a concession among member countries to reduce the tariffs on specific IT products to zero by a specific year. This is in addition to those already agreed upon under the GATT. Though free movement of IT products across the world is expected to bring down the cost of making such products by proliferation of global value chain and production network and by encouraging competition, the reality is that the major gain goes to those companies who control the technology. Indirectly, the entire manufacturing process and service industry receives a major boost by adopting ICT as it enhances productive efficiency. However, several developing countries including India find it challenging for the growth of domestic ICT industry.

India signed the ITA-I on March 25, 1997, covering 217 lines of India's IT products on which tariffs have been reduced to zero level in the following manner: on 95 lines by 2000, 4 lines in 2003, 2 lines in 2004, and 116 lines in 2005 (Table 10.3).

ITA-I was expected to bring up competition by reducing the cost of components. OEMs need to focus on continuous product diversification. However, competitive space has been narrow in many countries including in India as concentration has been high with significant market share of

Table 10.3 Product coverage under ITA-I in general terms

S. No.	General categories covered under ITA-I	Broad products
1	Computers	PCs, laptops, input/output units
2	Semiconductors	Transistors, integrated circuits, microprocessors, electronic micro-assemblies
3	Semiconductors manufacturing equipment	Encapsulation machines, inspection apparatuses
4	Telecom apparatus	Telephones, pagers, mobile phones, switching equipment
5	Instruments and apparatus	Cash registers, postage-franking machines, electronic calculators
6	Data storage media and software	Floppy disks, CDs, software in physical support
7	Parts and accessories	Parts and accessories to the other six main categories

Source: Market Access Division, WTO

few companies. Most verticals of electronics are dominated by a few companies, such as Telecom equipment sector is dominated by Alcatel-Lucent, Ericsson, Nokia, Cisco, Juniper, and Huawei; semiconductor sector by Intel, Samsung, Texas Instruments, Toshiba, AMD, ST, Microelectronics, and Analog Devices; mobile handsets by Apple, Samsung, Nokia, RIM, LG. It has acted as a huge barrier for the potential entrants in manufacturing or innovation process. In fact, at the initial stage, startup culture was not very popular, and venture capital and angel investors were largely absent in India. Due to the presence of few large companies in the world market, product design, prices, and even shaping up of the regulatory structure were dominated by them. As the technology has evolved very fast, many countries including India struggled to update its regulatory environment time to time. Many experts feel that too much involvement by big companies in controlling prices and pushing global standards as per their convenience hinders local development of technologies which is essential for addressing the digital divide. In other words, developing countries like India have been constrained to use electronic products catering to the tastes and preferences of the developed world. Thereby, further accentuating the digital divide.

In post-WTO regime, we have observed a significant decline of tariffs around the world, but the benefit is largely neutralized with the rising non-tariff barriers (NTBs) in the form of licenses, standards, strict regulatory regime, and so on (Mukherjee and Kapoor 2018). Product-specific standards are allowed under technical barriers to trade (TBT) measures of the WTO. The TBT structure in IT products inhibited India's exports to grow. On the contrary, absence of sound regulatory environment and independent standards allows free flow of imports. There has been a huge decline in investments in the manufacturing of components, raw materials, parts, and even electronic sub-assemblies. High-value-added manufacturing has been affected the most. However, India is now extensively working on domestic standards development.

In these 20 years, though the members of the ITA-I have increased significantly, the products covered have not changed at all, despite the IT sector having developed rapidly. Due to this, a sub-group of ITA-I members comprising the EU, the USA, and China embarked in 2012 on negotiations for an ITA-I expansion in terms of product coverage which were concluded successfully in 2015 and finally agreed upon at the 10th Ministerial Conference of the WTO in Nairobi in December 2015. This is popularly known as ITA-II. It is basically an effort to increase the coverage of IT products on which customs duty would be bound at zero, addressing non-tariff measures and expanding the members to include new countries such as Argentina, Brazil, and South Africa. ITA-II includes 50 members with an additional coverage of 201 products worth $1.3 trillion per year. The intensity of tariff rate reduction under ITA-II is faster than ITA-I as a large part of the tariff liberalizations occur instantaneously upon entry into force. The updated list of products covered under ITA-II includes a few consumer electronic items and certain security-related products.

10.18.2   E-commerce

Electronic commerce or e-commerce is the activity of buying or selling online. According to the World Trade Organization's (WTO) work program, electronic commerce involves the production, distribution, marketing, sale, or delivery of goods and services by electronic means. With the integration of the different economies, global e-commerce trade has seen a significant growth which is expected to continue in the future. It has been playing a vital role in various business activities. The USA and China are among the largest exporters of e-commerce, while India is among the fastest growing markets for e-commerce.

WTO members adopted the "Declaration on Global Electronic Commerce" in May 1998 to create new opportunities for trade in e-commerce. A comprehensive work program on e-commerce was established in September 1998 to examine the global e-commerce related trade issues and the development implications, e-commerce in the GATS framework, and intellectual property issues.[1] It was agreed that no custom duties will be imposed on electronic transmissions. Also known as "moratorium on customs duties", it has been renewed regularly at each ministerial conference. The bulk of this value covers the services.

India asked to maintain a balance between WTO members' right to regulate in the areas of domestic regulation, protection of privacy and public morals, and prevention of fraud.[2] It also highlighted the relevance of international standardization to e-commerce. Interconnectivity and interoperability of domestic information communications structures should be allowed as any mismatch would indirectly create monopolies and cartels in the global markets and would restrict the involvement of developing countries' firms.

Though India has seen fast growth of e-commerce, business to consumer e-commerce market is being driven by stiff price competition as online retailers like Amazon.com Incorporated, Snapdeal, and so on provide huge discounts and payments are largely made on a cash-on-delivery basis. Companies find it difficult to retain customers. India lags far behind China, Brazil, the USA, and the UK in key information and communication technology indicators.

This sector is widely discussed in many international organizations and multilateral forums. Many countries are now keen on having free trade rules in the e-commerce sector under the WTO. But, the developing countries need policies like subsidies, tax benefits to domestic firms, protection of infant industries, and the right to use proper value addition rules while importing. They have expressed concerns that WTO rules may decrease their ability to support domestic industry.[3]

India is engaged in bilateral and regional trade agreements like RCEP, and e-commerce is a key component of such agreements. However, India

---

[1] https://www.wto.org/English/tratop_E/ecom_e/ecom_e.htm.

[2] https://docs.wto.org/dol2fe/Pages/FE_Search/FE_S_S009-DP.aspx?language=E&CatalogueIdList= 16,320,20,805,43,530&CurrentCatalogueIdIndex=2&FullTextHash=371, 857,150&HasEnglishRecord=True&HasFrenchRecord=True&HasSpanishRecord=True.

[3] https://docs.wto.org/dol2fe/Pages/SS/directdoc.aspx?filename=q:/WT/MIN17/21.pdf.

has expressed concerns over participating in e-commerce discussions in the WTO. It submitted a formal document to the WTO in November 2017 opposing any negotiations on e-commerce.[4] It stated that it would continue to work under the WTO's work program but would not participate in any negotiations related to opening cross-border digital trade. India's domestic policies are still evolving, and these negotiations can badly affect the ability of the government to promote domestic industry.

Classification of digital products like music and e-books, and so on, as goods or services, and the extension of the moratorium on custom duties on electronic transmissions have still not reached a consensus.

Developed countries like the USA and EU have healthy regulations regarding consumer protection and data security. The same is missing in the case of developing countries. They are doubtful regarding their capabilities to implement new regulations at a pace which could keep up with the technological changes.

There is lack of information on e-commerce exports and imports, on how business works and on how WTO rules on e-commerce are going to affect Indian companies. This is the reason why India has taken a defensive position in the negotiations for e-commerce.

10.19   Conclusions

India had a long and complex history of regulations. Like other nations, it is struggling with its regulations to both cope with and to exploit the new technologies. In the past two years, it has made bold decisions in line with progressive thinking for improving the lot of its underdeveloped population and to propel the economy and its society into the new age on the world stage. However much remains to be done. As of the writing of this book, the regulations are clearly divided:

- Some are aimed at protecting the status quo, for example, banning mixed calls in order to protect the Telcos, banning bitcoins as their impact is currently unpredictable.
- Some are cumbersome and regressive, for example, heavy license fees on spectrum and tax levies on Telcos, no tax breaks for funding start-ups, no incentives for DCB, and so on.

[4]https://docs.wto.org/dol2fe/Pages/FE_Search/DDFDocuments/240274/q/Jobs/GC/153.pdf.

- Many are aggressively modern and opportunistic, for example, Net Neutrality, drones, AI, IoT, insolvency and bankruptcy code to de-risk investments in startups, and so on.

The sector, as well as most of the existing regulations, will undergo a positive transformation due to the new draft policy on National Digital Communications released by the Department of Telecommunications. What one hopes to see is an integrated policy framework run in a coherent, unified fashion by the different government departments. Currently, the approach is fragmented with different pieces being proposed/run by different departments like TRAI, DoT, MeitY, Ministry of Communications, and so on. It is proposed that ICT be under one ministry rather than it being split under two ministries. Overall, the effort is in the right direction, and much more progress is expected in the near future, especially with policies like NDCP 2018.

The Takeaway Box: India's Regulations Are Rapidly Evolving

- India's changing regulations point to a very dynamic situation in which the Indian government has a multi-pronged agenda: to leapfrog India into the digital economy, enable industrial growth, seize new global opportunities, and develop India domestically.
- Many of the regulatory changes made are beginning to show results. Some of them are in the pipeline, and others need to be made (as per some recommendations in this chapter).
- Corporations, industry bodies, NGOs, academia, experienced individuals, governments at every level (and their departments) need to contribute to help shape the regulations. Otherwise the representation will be skewed, and the policies will take a distorted shape.
- India is also actively engaged in discussion at a multilateral level to ensure that there is a level playing field among Indian companies and MNCs. It has opposed the move by some countries to further liberalize the trade regime related to ICT and e-commerce. India's regulatory regime and domestic industry is at the evolving stage, and hence it would like to go slow at least for the next few years.

References

Aadhar LPG Scam. (n.d.). Retrieved from http://www.business-standard.com/article/current-affairs/aadhaar-exposes-5-cr-ghost-accounts-3-5-cr-fake-lpg-connections-goyal-117110400648_1.html.

Aadhar Mid-day Meal Scam. (n.d.). Retrieved from http://www.hindustantimes.com/india-news/midday-meal-scheme-aadhaar-exposes-4-4-lakh-ghost-students-across-3-states/story-9Kna9AHXQYX59bn4bzvrzO.html.

Aadhar Scams. (n.d.). Retrieved from http://www.thehindu.com/news/national/aadhaar-helped-save-about-50000-crore-through-dbt-centre-tells-supreme-court/article18937378.ece.

Accenture. (2017, December). *Artificial Intelligence has Potential to Add $957 Billion to Indian Economy.* Retrieved from https://yourstory.com/2018/02/budget-2018-artificial-intelligence-fuel-indian-economy.

Afuah, A. (2003). *Innovation Management: Strategies, Implementation, and Profits.* Oxford: Oxford University Press. ISBN-10: 0195142306, ISBN-13: 978-0195142303.

AYUSH, Ministry of AYUSH, India. (n.d.). Retrieved from http://ayush.gov.in/.

Das, G. (2012, September). *India Grows at Night.* India: Penguin, ISBN-10: 0670084700, ISBN-13: 978-0670084708.

DNIS. (2013). *National Universal Electronics Access Policy, Disability News and Information Service.* Retrieved from https://www.dnis.org/National-Policy-on-Universal-Electronics.pdf.

Ghoshal, A. (2018, January). Government's Botnet Malware Apps Bring Down Infections in India. *VC Circle.* Retrieved from https://www.vccircle.com/govts-botnet-malware-apps-bring-down-infections-in-india-report/.

Juniper Research. (2018). Cybercrime Will Cost Businesses Over $2 Trillion by 2019. Retrieved from https://www.juniperresearch.com/press/press-releases/cybercrime-cost-businesses-over-2trillion.

Kak, A. (2017, November). *Net Neutrality: Trai Does Well To Recommend Hard-Coding It in Licence Terms.* Retrieved from http://www.business-standard.com/article/economy-policy/net-neutrality-trai-has-done-well-to-recommend-hard-coding-it-in-law-117112801376_1.html.

LAWS, Lethal Autonomous Weapon System. (n.d.). Retrieved from https://en.wikipedia.org/wiki/Lethal_autonomous_weapon.

MeitY IoT. (2016). *Ministry of Electronics and Information Technology.* Retrieved from http://meity.gov.in/content/internet-things.

Mukherjee, A., & Kapoor, A. (2018, March). *Trade Rules in E-Commerce: WTO and India.* Retrieved from http://icrier.org/pdf/Working_Paper_354.pdf.

National Skill Development Mission. (n.d.). Retrieved from http://www.skilldevelopment.gov.in/nationalskillmission.html.

NDSC, National Skill Development Corporation. (n.d.). Retrieved from http://innovation.nsdcindia.org.

NTP2018, TRAI (Telecom Regulatory Authority of India). (2018, February). Retrieved from www.trai.gov.in/sites/default/files/Recommendation_NTP_2018_02022018.pdf.

Parbat, K. (2018, February 1). Telecom Companies Disappointed With No Levy Cut in Budget. Retrieved from https://economictimes.indiatimes.com/tech/hardware/telecom-companies-disappointed-with-no-levy-cut-in-budget-2018/articleshow/62740042.cms.

PMJDY, Pradhan Mantri Jan Dhan Yojana. (n.d.). Retrieved from https://www.pmjdy.gov.in/account.

Regidi, A. (2017, December 3). *Understanding the Data Protection White Paper Part III: Defining 'Processing', Cross Border Data Flows and Data Localization*. Retrieved from http://www.firstpost.com/tech/news-analysis/understanding-the-data-protection-white-paper-part-iii-defining-processing-cross-border-data-flows-and-data-localization-4239419.html.

Russell, S. (n.d.). Retrieved from www.autonomousweapons.org.

Skill India. (n.d.). Retrieved from www.skilldevelopment.gov.in.

Telecomera. (2018). *Industry Take on Budget 2018*. Retrieved from http://telecomera.net/blog/industrys-take-on-budget-2018/.

TRA. (2018). *Regulation of Drones in India, TRA*. Retrieved from https://www.tralaw.in/regulation-of-drones-in-india/.

UN Urban Habitat Reports, UN Habitat's online reports, UN Habitat. (n.d.). Retrieved from http://urbandata.unhabitat.org/compare-cities/.

Watal Committee Report on Digital Payments. (2016, December). *Medium Term Recommendations to Strengthen Digital Payments Ecosystem*, Ministry of Finance. Retrieved from http://finance.du.ac.in/du-finance/uploads/pdf/Reports/watal_report271216.pdf.

CHAPTER 11

Winning the Competition

Know thy self, know thy enemy. A thousand battles,
a thousand victories.
—Sun Tzu

11.1 INTRODUCTION

With rampant uncertainty and increasing competition in the market, what should an organization do to thrive and sustain its position? Traditional wisdom suggests building a strong brand name with a unique value proposition at a competitive price. These are correct, no doubt, and still hold, however, the path to reaching this state is no longer conventional. The new routes must essentially include co-creation, innovation, customer-centricity, and design thinking. All organizations, especially the large ones, should build a new kind of organization to empower every employee and build intrapreneurship into their culture.

As described in the Introduction of this book, innovation can be explained through a new product or services which comprises of either new technological knowledge or new market knowledge, or both (Afuah 2003). And as mentioned in the chapter on ICT tsunami, ICT can drive economic value at the junction of innovation, strategy, and capability (i.e. resilience, skill adoption and entrepreneurship) of the organization. Thus, specialization, innovation, collaboration, and integration become quite important.

© The Author(s) 2019   313
S. Birudavolu, B. Nag, *Business Innovation and ICT Strategies*,
https://doi.org/10.1007/978-981-13-1675-3_11

Every organization must be viewed through the lens of ICT to exploit the transformational role, the magnifying power and the ability for fine-grained control, analysis and intelligence that ICT affords. It's good to gain an excellent understanding of technology, especially the key technologies listed in the chapter on technology and enterprise landscape, in order to know how to leverage them.

Establishing a strong business focus is extremely important as discussed in the chapter on focus, which also describes the processes and framework for developing and retaining focus. The chapter on strategy describes the entire process and framework of establishing an effective innovation strategy for the organization. Without a critical mass of social capital in the organization, all strategies and investments are doomed to fail. The chapter on social capital details the ways to build solid social capital in the organization; it is the most crucial asset of any organization.

The nuances pertaining to regional differences and the know-how about how to exploit these for launching product/service in a different geography are vital attributes for any growing, global organization, and the chapter on regional factors addresses precisely that.

The two chapters on India explain the current scene of ICT and the ICT regulatory structure in India. There is much to learn from a vast, developing, diverse, democratic, and progressive country like India.

With all this background, an organization is in a great position to reach success. But what does it really take to clinch the deal and win the competition in the market?

Enterprise architecture and design thinking deserve priority in the vision and strategy of an organization. Design Management Institute (dmi.org) that has constructed a DVI (Design Value Index) has hand-picked 16 US stocks of companies that are very "design-centric" in their approach, as measured on 6 criteria. The DVI stocks have beaten the Standards & Poor's 500 stock benchmark by 211% (DVI), clearly indicating that the design approach has significant untapped potential in the industry. Agility is another cornerstone for success in the new age. Several industry behemoths have been overtaken by the new players, not because they lacked technology or the money/resources, but because they lacked speed in their business, that is, in creating new models, in seizing opportunities, and in implementation, execution, and operations. They ceased to think like entrepreneurs and were more focused on executing and extending their entrenched models.

The key question therefore is, how do you create, realize, and capture value?

Clearly there is no single answer that will fit all organizations, but there are patterns that one can and must exploit. This chapter explores them.

11.2   FROM PRODUCT-CENTRIC TO EXPERIENCE-CENTRIC MINDSET

The entire market has moved from being product-centric to being experience-centric. Cloud computing is a good example of this. Customers don't really want the software, they only want the benefits of the software. So, they use the software through an interface, which provides them with an experience without having to deal directly with the software. The experience could be whatever is desired or expected, such as mail, social media, video streaming, research on a topic, comparison of several customer goods, and so on.

It is a shift in mindset. Products provide the basis around which to build services. A cohesive set of services provide an experience. And a superior experience improves customer stickiness.

A product, by definition, encompasses core functionality that can be used by many. This is in contrast to a solution, which is crafted for a specific use (or for use by one client). A product solves the problem for many, while reusing the implementation, which is kept as minimal as possible. Product management involves extracting the common requirements from discussions with a variety of customers in order to come up with the product feature set that will be used to build the product. There will always be customer-specific requirements that cannot be accommodated in the product because they will either mess up the design of the product or are of no interest to the other customers. These customizations will need to be done as a part of solutions that are based on the product. And it is these tailored solutions which ultimately provide the exact services needed by an individual customer. The complexity and cycle-time to make these customizations needs to be drastically cut short for the product/solution/service to be successful. A platform allows customizations to be done seamlessly and through standard templates.

The final goal of a product is to create top class customer experiences that are orchestrated and supported by services which provide high value to its customers. To accomplish this, one needs strong product management and needs also to use data and information coming out of operations, that is, using the product in production with live customers. It is a positive feedback loop, in which the data from operations is analyzed and is used to innovate and improve the product. This leads to a superior customer experience, which in turn attracts more customers and also improves customer loyalty. This yields richer and more sophisticated data from operations, which is extremely useful for innovation. And innovation leads to new and better kinds of products and services. For providing a superior customer experience, it is necessary to build customer stories and narratives, rather than merely jump to listing out a set of product features.

The next goal then is to co-create services through open innovation by collaborating with other players in the ecosystem. Again, this is best done through platforms.

11.3   Adopt Design Thinking (or Rather "Design Doing")

Design thinking is about borrowing creative strategies that designers use (during the design process) in the business or social context. "Design doing" is better than "design thinking" because design is the process of hacking and prototyping. A good design is achieved after several iterations. Mere thinking about design usually doesn't go anywhere; it can lead to analysis paralysis. However, we'll go with the phrase design thinking as it is in vogue.

Today, all the industry leaders—such as Apple, Samsung, Google—use design thinking. D.School, which is Stanford's Hasso-Plattner Institute of Design (D.School), has identified five phases for design thinking:

1. *Empathize* with your users, to understand their experiences and motivations, and get immersed in their environment. Get a deep personal understanding of the human needs.
2. *Define* the problems/needs of your users and your insights. Collate and analyze information from the empathize phase and formulate the core problems in a concise problem statement. The problem should be framed in a human-centric way.
3. *Ideate*, by challenging assumptions and creating innovative solutions. Use ideation techniques to get as many solution ideas as possible for the problem.

4. *Prototype,* By creating small-scale experimental products that show-case the key features of the solutions produced in the previous stage (Ideate). Examine each of the prototypes from the point of view of user experience and improve/accept/reject them. This phase and the next demand a hands-on approach.
5. *Test* the product thoroughly against the solution features. Use the results of the testing to redefine the problem and refine the previous phases. It is far better to encounter failures early and refine the product through iterations rather than continuing to heavily invest time, effort, and money into building the main product and failing in major, unexpected ways at a later stage. The refinements should be done as early in the game as possible.

There are feedback loops between phases, hence, the process is not strictly linear, but is in fact non-linear. Design thinking arms everyone with the tools to apply the methodology to solve complex problems in all walks of life, in the face of uncertainty and instability. It is very essential to build design thinking in every area and into the culture of the organization.

As per Matthew Kressy, director of IDM (Integrated Design and Management) Lab of MIT (Massachusetts Institute of Technology), success in today's world for an organization requires an intelligent combination of the following three (MITidm 2018):

- Business (indicating viability)
- Design (indicating desirability)
- Technology or engineering (indicating feasibility)

Each of these three functions talk a different language. Hence, all teams should be built as individual siloes, each with all the three kinds of people in it: business people, designers, and engineers. Then, each silo can talk to another silo without losing stuff in translation. Do not build separate teams each for business, design, and technology. They may not be able to talk to each other properly. Design is typically targeted at the users; hence, their involvement and feedback should be taken throughout the cycle.

As per IDM's philosophy:

Sustainability should always be a top goal; wastage should be reduced. However, the real opportunity is in increasing value to the customers, not cutting costs. This is done by making Design a priority, to increase desirability. Hence hire designers. This is a huge opportunity wasted around the world. Design resonates emotionally. Many products or services are bought with emotional filters switched on, such as "does this make me feel good or look good". Hence it is best to sell experience, not a product. Most of the products that command premium value (and make high margins on their product) have an emotional value attached. Over a period of time, when you develop a brand, you can develop leverage brand, i.e. have brand equity. When you have brand equity, you can leverage that into better products and make more money.

To build a good organization, you need a good culture, with great people. Mere skills are not sufficient, you need to have people who put team first, and put excellence first, and are compassionate. This builds a vibrant organization.

A good culture needs a great environment. Design process is inherently messy, as you keep trying out different things. Experimentation is an inherent part of a great organizational culture. The culture should be collaborative and open, should encourage creativity, and should be fun. The team should be well integrated and have a shared vision. Hire the people who you want to run the world.

The key steps in the design process are:

- Explore—find out what's wanted
- Create—make what's wanted
- Implement—make lots of what's wanted

There will be many iterations between the first two steps. This resonates with the Plan-Build-Measure-Learn process prescribed in the Lean Startup Model (Ries 2011).

About leadership: good leadership is easy to achieve if one follows a design process and also integrates the team well. To extract requirements from the customers, you can use a variety of methods, such as mapping customer journeys, storytelling, interviews, focus groups, and so on. It is always essential to talk to the stakeholders to get a real feel of their problems and perspectives.

The leader can be biased owing to his background. If he has a technology background, there is a tendency to give less importance to business and design. If the leader comes with business experience, he may overemphasize finance at the expense of technology and design. A leader who is a designer may produce products that are aesthetically very appealing but have limited functionality and which struggle for business viability. It is important that all the three are well balanced. Therefore, by understanding one's own biases and shortfalls, a leader should take special care to compensate for those areas, or the balance will be lost.

Push the limits and explore the boundaries. Making mistakes should be quite ok. Emotions are integral for getting a real understanding. Understand that all technology is eventually going to get commoditized, so there will be no great differentiator there over time. It is best if the entire team empathetically hears out the customers, especially the users to gain real insights.

11.4 Automation Is Key

To produce and deploy services most rapidly, automate wherever possible. The general principle is: *Do not do things but build systems that do things.* To get the best out of automation, adopt a mindset where automation is considered the norm and human intervention is treated as an exception. Algorithms are getting superior by the day. By default, put the algorithm in control rather than the human being, and keep tweaking for even better results. If you have manual intervention as default, then switching to automation is a major chore, rife with all sorts of problems, including psychological barriers. Barring a few highly complex environments that are full of Knightian uncertainty (Knight), that is, risk which can't be measured, for example, strategic decision making, most environments are amenable to automation because a lot of data can be gathered. Even many of the highly complex environments can also stand automation, for example, autonomous driving vehicles, robotic surgery, flying airplanes and spacecrafts, and so on, because it is possible to measure data and calculate risk and uncertainty.

Examine the entire lifecycle, re-engineer the processes, and automate them. All the critical parts (relationships, activities, use-cases) must be automated. In fact, by definition, a product itself is an automated, reusable asset that has a compact implementation of the core use-cases, that is, common patterns. Processes must also be automated. An automated system yields benefits that are impossible to get otherwise, such as improving efficiency multi-fold, eliminating human error, ability to scale by orders of magnitude seamlessly, ability to run 24/7, integrating with other systems, function in layers, and producing operational data continuously.

In the current day, some of the key factors for automation are usage of both algorithms and data, high connectivity, standardization, cloud computing, microservices, mobility, AI/ML/robotics, decentralization, and platforms.

It is important to use standard frameworks such as containers implementing microservices. The automation should be easy to maintain, and also integrate well with other components. Standards should include industry (business) standards, regulatory, and technology standards.

Automation uses both algorithms and data. Data analytics, mining, AI, and Machine Learning lean heavily on data to identify patterns, adjust behavior in real time, and continuously improve accuracy and performance. Industrial automation and control systems rely on feedback loops.

11.5   BUILD FOR SCALE

Once a business model and its related product are past the initial prototyping and proving stage, then growth is a necessity for the business. Growth is either powered by infusion of capital to scale or through other means, such as viral marketing through social media and rapid adoption. It is hard to retrofit scalability into the built product, if the product is a success. Typically, one loses momentum (and the market opportunity) if one has to go back and rebuild the product to handle scale. Many organizations (large and small) have experienced this painfully. Scalability needs to be an inbuilt attribute when the product is launched. What about reliability and availability? These are already assumed to be part of the quality parameters of the product, because there are SLA expectations, whether official and paid (as in commercial services) or informal and unpaid (as in social media apps and emails). By default, the new standards are 24/7 availability (i.e. an uptime of 5-Nines, i.e. 99.999%) and 100% reliability in all transactions.

Note that a business model implementation includes both the product and the processes. Both need to be scalable. The scalability should be nonlinear to get the maximum leverage. The idea is that in order to get five times the work done, one should not have to invest five times the resources (which include people, time, money, material, computing resources, etc.). In fact, the architecture, design, and implementation should be such that with every unit of additional investment, the output should be disproportionately great. For example, if initially one person can do the work of ten people, five people should be able to accomplish the work of five hundred.

Much has been made of the exponential growth model, which suggests that companies should work towards building businesses that grow exponentially, not linearly. This is good, and one must test out the business model at an early stage. If it is indeed catching on, then there is a case for "viral" growth. However, it should be noted that exponential growth constitutes one part of the product lifecycle and can't last forever. If perpetual exponential growth were possible for all companies, then any company would have the entire population of the earth as their customers within a decade or two. This is not true of even the market leaders, because they face competition from many quarters, thereby restricting indefinite growth. Clearly only a handful of companies can do that and become global leaders. Even those organizations would have to play a very deep, long-term game, investing heavily in several areas without hope of profits for years on end, for example, Amazon, several divisions of Google, Microsoft, and so on. A lot of organizations fall in the long tail and are better suited for overall linear growth rather than exponential growth.

One must build for scale because it is possible that the initial product version may enjoy exponential growth during part of its product lifecycle before peaking off or even declining. If it reaches a steady state at some stage, it would still have to cater to the demands of its full acquired customer base, without missing a beat. It is also possible while the product is reaching the peak of its lifecycle, they could pivot the business model and release a new version of the product for the market. And so, a new product lifecycle commences from that point onwards, thereby entering another phase of exponential growth.

Organizations must get comfortable with floating new business models rapidly—and also retiring them off—either when they're not working or when they've reached the end of their lifecycle. Acquiring, retaining, and migrating customers to new models should be seamlessly done.

All these indicate that that the bulk of the underlying "machinery" and processes must be standardized and automated, to enable jugglery in the higher layers, that is, flexibility, nimbleness, and scalability. And this is what a platform is all about.

11.6   What's Your Platform?

The concept of a platform needs to be understood clearly, because platforms rule the economy. Platforms are key to success in the digital economy. In fact, one absolutely cannot hope to win in today's competitive marketplace without leveraging platforms. This is needed for innovation, scale, flexibility, customer experience, and sheer stickiness with all the ecosystem partners, not just customers. Platforms help in solving problems for the multitude. Different products are needed to suit different customers and markets, regardless of the underlying (unified) architecture of the common base product. As there cannot be a one-size-fits-all approach to address the needs of the multitude, it is necessary to open up the game to allow different possibilities, permutations and combinations, and entirely new solutions to the extent that the framework will permit. And this is what a platform aims to accomplish. The customer layer allows entirely different business models to be built. Tweaking of the business models should be easy and fast, in order to fine-tune to the customer needs.

All the leaders like Amazon, Google, Uber, Facebook, IBM, GE, and eBay have built highly successful platforms.

A platform can be described in many ways:

- Platforms separate out the logic functions of applications in a layer, so that an IT structure can be built for change (Brigwater 2015).
- A platform is a group of technologies that are used as a base upon which other applications, processes, or technologies are developed (Techopedia).
- The platform conforms to a set of standards that enable software developers to develop software applications for the platform (Techopedia).
- Platforms are structures that allow multiple products to be built within the same technical framework (Clarks 2009).
- Platform is a business framework that allows multiple business models to be built and supported (Brigwater 2015).
- A platform is a business model that creates value by facilitating exchanges between two or more interdependent groups, usually consumers and producers (Moazed 2016).

Thus, a platform enables and catalyzes the creation, orchestration, and distribution of value. So, the platform is the underlying "machinery" abstracted out as a layer, on top of which different use-cases can be built, interfacing with the platform in defined ways. Hence, the platform itself exists for longer and changes more slowly than the business models that are built on top of it.

A platform is not an individual chunk of technology or a single product or service. It is a framework that allows new products or services to be built easily and reduce transaction costs (due to reuse and speed) and also enable open innovation. External parties can combine their innovations with the features and power provided by the platform to build new products and services. Most SOA (software-oriented architectures) and Cloud Services attempt to provide platforms, for example, SaaS, PaaS, KaaS, IaaS, and so on. Platforms create communities and markets with network effects that allow users to interact and transact (Moazed 2016).

The initiative named Platform Hunt has described nine types of software platforms (Platform Hunt), which include platforms pertaining to technology, computing, utility, interaction networks, marketplaces, on-demand service, content crowdsourcing, data harvesting, and content distribution

Let's understand these briefly:

1. *Technology Platform*: These provide technology infrastructure and are invisible to the end-user. There are no network effects or multi-sided aspects as regards business. The software developers use these to build products. Examples are Google Cloud, AWS, and Oracle Cloud. Technology platforms are the simplest.

2. *Computing Platforms*: These provide an interface between two groups: the developers of applications and end-users of the platform. Apple iTunes and Google Play Store are good examples. The developers then have a ready platform with a mass customer base that is willing to try and use their applications. The users have a convenient and safe marketplace to go to for a variety of needs from entertainment to business and personal productivity. There are strong network effects, leading to explosive growth when the platform takes off. These in turn spur the adoption of the general ecosystem, for example, devices such as smartphones, PCs, and gaming and other tools.

3. *Utility Platforms*: These provide a "freemium" model. The bulk of the users use the free service and a handful sign up for paid services. The latter grows in terms of value provided and the inflow of revenues and can attract enterprises as well for premium business services with a different level of quality, for example, with SLAs, type of offerings, and so on. Another use is that the free subscribers can themselves be targeted for premium services or for advertising.

Examples are Google Search, Gmail, and PolicyBazaar.com which provide free services but use the user data for promoting targeted advertising and collect revenue from the advertisers.

4. *Interaction Networks*: Everything revolves around the user's identity on this platform. The user's preferences, habits, social and business network, and so on are key. This truly uses network effects because there is a multiplier effect as the number of users on the platform increases and the users start interacting with one another. Examples are LinkedIn, Facebook, WhatsApp, Pinterest, and so on.

5. *Marketplaces*: These provide two-sided platforms which bring together buyers and sellers. The network effects are due to a growing positive spiral in which an increasing number of buyers leads to an increasing number of sellers, which in turn attracts more buyers to the platform. Examples are eBay, Amazon, Alibaba, and so on.

6. *On-Demand Service Platform*: These are focused on delivering end-to-end services in a specific market. The services are measured by user experience SLAs and quality metrics of different kinds, so the performance measurements are stringent. Examples are Uber Taxi, Swiggy, CallHealth, and so on. A large number of providers, for example, the number of taxis on Uber or the number of food vendors on Swiggy, increase the chances of pulling greater number of consumers towards the platform. Some peer-to-peer platforms also fall into this category. For example, ATL (Any Time Loan) connects the lenders to the borrowers one-to-one for dealing out micro-loans, thereby matching the needs to the supply.

7. *Content Crowdsourcing Platforms*: The users of this platform are prosumers, that is, both producers and consumers. They produce content and consume it. Examples are Wikipedia, YouTube, TripAdvisor, and so on. Typically, a portion of the user base produces content, and everyone is free to consume it. This is a great platform for people to express themselves, advertise their talents, pitch sales, educate others, spread their views, entertain others, and so on.

8. *Data Harvesting Platforms*: These platforms are data-centric rather than user-centric. The intention is to collect as much data as possible based on users' usage of the platform. In fact, the users sign up to provide and use data. Examples are IMDb for movie ratings and summaries, mouthshut.com for customer ratings of goods/services, Google Maps for traffic prediction, Kaggle for predictive modeling

and analytics competitions, and Ushahidi for social activism and public accountability.

9. *Content Distribution Platforms*: These platforms function as intermediaries which bring together content owners to specific users desiring that content. The platforms have their touchpoints with individual users, whom they track, and connect the relevant content owners to the specific users. Examples include Google Ad Sense, Taboola, and Curata.

The key elements of a platform then are:

1. *Purpose*: This is the most important of all the elements. As described above, there are different kinds of platforms, and it is important to define upfront the purpose, the kinds of users, the use-cases, user journeys, and requirements, including the non-functional requirements, like performance and scalability.

2. *Underlying infrastructure*: This refers to the platform's underlying machinery and inner workings. It consists of all the layers underneath the layer that it chooses to expose to the outside world. For example, this could be hardware, software, and basic units of functionality/service that can be used as building blocks. Some of the commonly expected features could be security, scalability, and ability to integrate with/among the blocks seamlessly.

3. *Layer of Abstraction* that shields the inner workings of the infrastructure. The key thing here is the modeling of the platform. This includes the different components of the platform, their interrelationship, the functionality provided aligned to the purpose.

4. *APIs (Application Program Interface) or Web Interface/User Interface* in the platform that allows the external systems and parties to interface with the platform in defined ways.

5. *Standards (Industry, Regulatory, Technology)* that the platform adheres to. This enables ecosystem players to plug into the platform without having to develop platform-specific custom interfaces that are brittle and are expensive to maintain.

6. *Communities*: The different interested parties that the platform brings together to bring about synergies. The larger the community, the stronger the links, the greater the interactions, the more successful the platform is.

7. *Administrative interface* that allows reconfiguration and customization of the platform itself for meeting the needs and demands of the business models built on top of the platform.

Telecom companies are setting up labs with platforms to enable different vendor-partners to experiment and launch services on the platform to the subscriber base of the telco (Analysys Mason 2015). The services vendors may tweak their business and technical models to arrive at a valuable offering that jells well with the customers. On the other hand, if a model doesn't work out, they can discard it fast without incurring too much cost. The feedback loop is rapid, and it is a more efficient way of unlocking value.

11.7   The Rise of Metadata

It is estimated that 80% of the underlying routine IT infrastructure operations will be automated by 2020, and also over 50% of all enterprise data will be managed autonomously (Zeichick 2018). As the underlying layers get automated, commoditized, and even outsourced, there will be low, if any, competitive advantage left in running marginally better operations. As the layers keep maturing, there will be progressive commoditization in layers above also. The tools, building blocks, and frameworks will get more sophisticated—bigger, fine-grained, complex, assorted, pre-built, ready-to-integrate, and so on.

However, the real deal will be the way the different models will be built and used to deliver value. And the crux of entire business operations and strategy lies in the data pertaining to the customers and their transactions. As the data is itself a complex ocean, the key lies in the metadata—or the data about the data. These are the deep insights gleaned from the data through various means—either gathering them directly, through surveys, data mining, big data analytics, small data analytics, mapping user stories, case studies, and so on. The real competitive advantage lies in how one gathers and uses the data from all the stakeholders—customers, vendors, partners, end-users, one's own organization, developers, and administrators.

For example, the real value of Google Search is not that it searches and provides information, but that it uses metadata about which site is visited more often (i.e. Page Rank Algorithm) and presents the search results in that order (highest popularity websites first). Amazon shows user reviews and ratings for books/products, as well as suggests to the user what the related popular products are. The same goes for all other services, from movies (Netflix) to ordering online food. The users can in fact drill down to individual feedback and review comments from other users, to decide what is relevant to their own individual needs.

11.8   What Are Your Communities?

As is clear from the section on platforms, communities are key to the success of a platform. There are different kinds of communities depending on the platform, for example:

- The end-users: the main users of the platform
- The developers: the developers who build applications on top of the platform
- The partners: the partner network on the platform
- The vendors: the vendors who sell through the platform
- The operators: the people who configure and operate the platform

Every organization must nurture its communities carefully. It should set up chatrooms, mailing lists, websites, conferences, meetings, trainings, and industry connects to engage with its communities. Feedback and discussions within communities are worth their weight in gold. Every product must have its platform and a community following, or it won't last for long in the face of competition. The organization must set up an entire team to manage communities, and this will be key to holding the market, innovating, and exceeding expectations.

11.9   Alliances Framework

In the new organization, alliances is a dedicated function. Alliances and channel partners not only empower the organization with a strong go-to-market strategy but also help in connecting intimately with the ecosystem, to gain intelligence and create value. Thus, alliances has a net value creation role, and helps one focus better on the core area.

An alliances framework is depicted in Fig. 11.1. There are four major components to an alliances function, which help create and drive value in three areas. This function needs to be a conscious effort driven from inception and from the top.

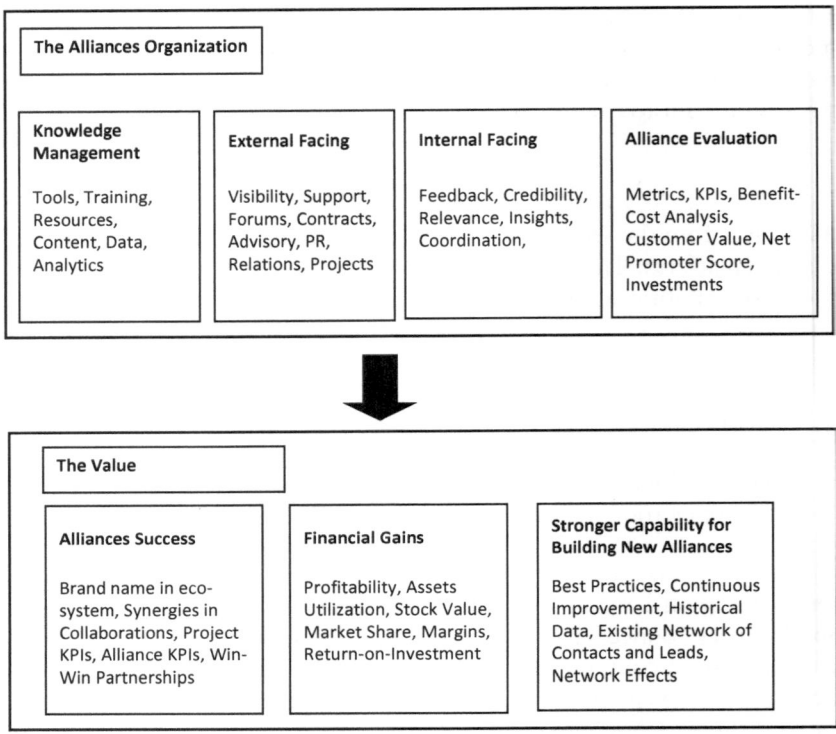

Fig. 11.1 The alliances framework (Source: adapted from Roberts 2002)

In open innovation, alliances are paramount. The entire alliance needs to be extremely well orchestrated from inception, through the various cycles, and until closure. An alliance can be closed if it is no longer relevant, is not yielding much value, or is ending due to mergers and acquisitions. Alliances can be recreated, renewed, or resurrected in a new avatar, form, or contract for a different phase or business reason, even after being closed for a long time. For example, if a new technology or business model arises where it makes sense to reach out to the erstwhile partners, then it is time to reactivate a fresh engagement without any delay. The lifecycle of an alliance needs to be defined and tracked.

A basic checklist for the different stages in the alliance lifecycle are given in Table 11.1.

Table 11.1 Alliances checklist

S. No.	Stage in alliance lifecycle	Checklist
1.	Business case	Value chains, Requirements Engineering, make-or-buy decisions, collaboration criteria
2.	Partner selection	Screening and selection forms, technology roadmaps, patent domain maps, cultural fit
3.	Alliance formation	Negotiations matrix, prioritized requirements, contract template, alliance framework guidelines, alliance metrics
4.	Alliance governance	Issue tracking, reviews, alliance CRM, alliance communication plan, project plan, program management, contracts, change requests
5.	Alliance evaluation and closure	Metrics, surveys, feedback, periodic status reports (weekly, monthly, quarterly, annual), termination checklist, alliance closure

Source: Adapted from Roberts (2002)

An organization's alliances should be managed thoroughly. There are opportunities to make an alliance function in more ways than one, that is, not just to meet contractual obligations but also in other terms, such as sharing of experiences and knowledge, identifying new opportunities for collaboration, finding market access, and building templates, tools, and products.

11.10   EMBRACE LEAN AND AGILE THINKING

Lean thinking and agility go together. Speed is of essence, not technology or resources. In fact, technology and resources are meant to bring agility in delivery. Whereas intellectual property and technology can also be borrowed, ultimately it is important to deliver value *rapidly* to customers. For instance, the hugely popular Netflix runs on borrowed technology of AWS (Amazon Web Services).

Lean thinking focuses on maximizing customer value while eliminating wastage at the same time. It revolves around five principles:

1. *Value*: It's defined by the customer's needs, which include expectations, features, price points, and schedules.
2. *Value Stream*: This is the entire delivery process, from taking input to delivering the product/solution's service to the customer. It is

process re-engineering and identifies all the workflows and eliminates all unnecessary/wasteful steps in the process, resulting in an optimal delivery process.

3. *Flow*: This refers to streamlining the entire process after removing inefficiencies. And it results in a smooth flowing, optimal process.
4. *Pull*: The principle here is that it is the customer who should be in charge, and the process should allow the customer to pull value whenever he/she desires.
5. *Perfection*: This refers to instilling lean thinking in the culture of the organization and building social capital. It is only after repeated iterations that the principles sink in across the rank and file of the organization.

To make lean thinking successful, Kaizen, or continuous improvement of the value stream should be conducted through the PDCA cycle as follows:

- *Plan*: have a *hypothesis* and create a *plan* for building the product or the features
- *Do*: *build* the product/feature
- *Check*: *measure* and *learn* how the product functions or operates in production, under a variety of circumstances
- *Act*: *implement* and *deliver* the changes in the product or process, on a larger scale

These are also similar to the four steps of the lean startup model (Ries 2011):

Plan—Build—Measure—Learn

In the lean startup model, there is a follow-on step that has two options: persevere or pivot. If the first four steps yield the desired positive result, then you can press on with the model. Alternately, if the hypothesis is falling apart, then a course correction may be needed, and the idea needs to be refined based on the feedback from earlier steps.

The simple concepts of lean thinking and Kaizen have indeed revolutionized many industries. They are powerful and are being applied to the ICT as well—for example, incremental, iterative software development lifecycle through agile methodologies, rather than the bulky old waterfall model of development and prototyping, building, launching, and continuous improvement and delivery cycles in short sprints, rather than waiting for perfect Requirements Engineering and launching new releases per year.

The concept of agile thinking is meant to bring speed and flexibility into the process. The Agile Organization's definition is as follows (Agile Organization):

> Agile working is about bringing people, processes, connectivity and technology, time and place together to find the most appropriate and effective way of working to carry out a particular task. It is working within guidelines (of the task) but without boundaries (of how you achieve it).

There are minimal rules or constraints for agile working. The processes are kept light in the interests of speed and flexibility. In fact, many organizations are even sacrificing the heavy business requirements phase in favor of implementation. The new catch-phrases are "implement first", "pilot first", "hack it", rather than going through an elaborate phase of gathering business requirements and building workflows. This approach has been endorsed by many leading companies as a two-year study shows (Andriole 2018).

In ICT, agile software development methodologies such as scrum lay down only a handful of rules. To promote speed, there are sprints or definite regular delivery cycles, each consisting of a preset number of weeks (usually 6 weeks). The full (and changing) feature set or update to be delivered is known as the product backlog. In each sprint, a critical subset of the features from the product backlog is identified based on the customers' priority. Typically, small teams are constructed, each of which runs the project through sprints. The development and operations are fused into DevOps.

Agile thinking needs to be adopted across the length and breadth of the organization, not just by the development teams alone. It is pointless to do agile if the sales teams (say) are not following agile, as it would lead to mismatches between different parts of the organization and consequent wastage. For example, if the sales team has unreasonable expectations than can be accommodated in the sprint cycle, or if the agile scrum deliveries are not picked up by the customer with the delivery cycles, then the end-to-end process is quite wasteful and futile.

Similarly, the technology stack and the development and operations tools must support agile thinking. For example, cloud computing and containers with microservices are a great fit for agile processes.

It is essential to achieve excellent success at one level before moving on to conquer another level. This could mean either success in a geographic region, or with a specific customer group, or for a specific set of features, or with specific partners, or on a specific platform.

Use the traction gap framework (Wildcat Ventures 2017) to track and measure the progress and success of the product in the market. This framework has been described briefly in the chapter on strategy.

11.11   THINK OPEN INNOVATION AND CO-CREATION

The key to thrive amidst competition is to continuously disrupt yourself through innovation and continuous improvement. Bill Joy has stated that the smartest people in the world don't work for you (Joy). Hence, it is important that we leverage the talent, expertise, and experience of the world outside our own organization. With digital innovation platforms organizations can seize opportunities in a variety of markets:

- Offering own products and services (e.g. the telco's own products and services offerings)
- Offering other people's products and services (e.g. offerings from the telco's partners)
- Offering products and services co-created (e.g. telco's pooling resources for building open-source frameworks such as Linux Foundation for Networking (LFN), Telecom Infra Project (TIP 2018), and Open Compute Project (OCP)
- Targeting own customers in own markets (e.g. telco's own subscriber base for the services launched by the telco and its partners)
- Targeting other customers in other markets (e.g. other telcos' subscriber base for products and services not launched by other telcos)
- End-users (e.g. personalized services based on the data collected about them, social networks, communities, blogs, co-creation such as crowdsourcing data, etc.)
- Developers and entrepreneurs (e.g. toolkits, frameworks, templates and building blocks such as cloud platforms, ready-to-use payment-gateways, website building frameworks, access to subscriber-base, etc.)

While the examples given above are those for telcos, the same principles apply to all the organizations in all sectors.

This topic has been well explored in the other chapters of this book.

11.12   Be Competition-Centric

How does an organization know that it is on the right track and even be able to measure it? The best way is to be competition-centric as much as being customer-centric. If you're merely customer-centric, then it is likely that the competition has moved fast on something that your people have entirely missed. Then, despite your existing base of clients, your technology, your relationships with the customers, or your deep pockets, your customers may simply switch to over to your competitors because of such a compelling value proposition that the customers cannot ignore. Many a time, this happens because an organization's existing strategy, structures, incentives, and policies are so deeply entrenched that despite knowing about the rising new competition, the organization is unable to make the required changes in time to be able to survive. This has happened with Kodak which missed the digital revolution by a long shot, and with many other older behemoths whom the new ICT driven businesses have replaced. A comparison between the Fortune 500 list of companies 25 years ago and today would show the stark reality. It is a cliché to emphasize that it is essential to set up teams that do a deep competitive analysis and study the existing markets and technological forces to feed into the strategy. What is new is that this needs to happen in its own manner at every level of the organization. Every touch-point at every level, with the customers, partners, and other stakeholders, reveals something about the outside world and it must feed back into the organization. For instance, a simple revelation by the new college hires that their classmates who are good in AI have been hired into a certain competitor's team can indicate new moves by the competitor, or while interviewing candidates from another company we can learn about the tools and technologies that the other company employs.

It is very good to study the startups in one's line of business and the relevant areas that are problematic, to find entirely new approaches and solutions. One can collaborate with the startups that have innovative solutions or acquire them. In quite a few cases, the startups are acquired, not for the product that they've built, but for the brilliant team that could conceive such a product. The team can be used to rebuild a similar product from scratch or to repurpose it, to fit the organizational goals and customer base better. To scan the market for startups matching one's areas (or the problems for which one is seeking a solution), it is good to tie up with incubators, accelerators, venture capitalists, academia, and other players in the startup ecosystem.

Usually, any acquisition by a competitor is a red alert that serious changes are shaping up in the market.

It is also extremely necessary to have an enterprise architecture that allows the organization to be nimble and scalable to be able to respond quickly in the competitive market.

11.13   BUILD SOCIAL CAPITAL

To remain relevant and competitive in today's markets, an organization must espouse ambidextrous thinking and intrapreneurship. This has been elaborated in other chapters. While entrepreneurship is a no-brainer in a startup, the larger companies must inculcate intrapreneurship in every team.

Build an organizational culture with all the elements described in this chapter:

Services-centric mindset, design thinking, making automation a priority, building a platform and rallying people around it to form communities, building for scale, using data (especially metadata), embracing lean and agile thinking, and driving open innovation and co-creation models through strategy (not by accident), and finally, a competition-centric mindset keep the organization sharp and relevant.

Impart training and mentoring in all these areas immediately. Over iterations, morph the organization's culture and structures to align with these.

Obviously, the older command-and-control military type of organizational structures will no longer function in the new competitive world. Instead of hierarchical armies, run by rules and commands, visualize an organization as groups of commandos—highly flexible and extremely effective, and bound by a mission. Empower them with the relevant tools and training. Take in people who fit in with the organizational culture.

11.14   TELCO'S TRANSFORMATION JOURNEY

We can find corroboration of all the concepts discussed so far in this book, when we apply them to a telco, and discover that the steps map closely to the recommendations of the leading ICT industry experts. If Telcos were to seriously adopt the ten transformation journeys suggested by (Creaner 2017), then the Telcos would be on their way to true transformation (Table 11.2).

Table 11.2 Telco's transformation journeys mapped

#	Telco's suggested journey	Mapped to the concepts
1	From discrete network elements to an autonomous virtualized communications and cloud infrastructure	Digitalization, automation, loosely coupled, integrated architecture, cloud, platform, design thinking, technology, regulations
2	From reactive product-specific security to uniformly orchestrated security	From product-centric to experience-centric powered by services, digital security
3	From limited data exploitation to a uniformly orchestrated data-centric enterprise	Digital platforms, data analytics, information-centric, metadata, focus
4	From a closed management infrastructure to a platform for open services	Open innovation, platform, scale, traction gap framework, design thinking
5	From a limited portfolio of own services to managing a diverse portfolio of own and third-party services	Open innovation, quadruple helix, communities, intrapreneurship, innovation matrix, business model canvas
6	From a limited set of supplier relationships to an open ecosystem of partnership relationships	Open innovation, strategy, social capital, collaborative forums, regional factors
7	From a limited set of telco business models to multiple value creation and capture approaches	Open innovation, co-creation, value engineering, business model canvas, revenue-sharing, ambidextrous thinking, regional factors
8	From a vertical silo to flexible organization, culture, and operations	Platform strategy, social capital, platform, agile and lean operations, regulations
9	From focusing on own channels to market to adopting multiple channels for different vertical markets	Quadruple helix, innovation strategy, open innovation, ambidextrous thinking, forums, regulations and policy, regional factors
10	From one-dimensional management of customer relationships to 360-degree omni-channel management of the whole customer experience	Social capital, customer experience-centric, competition-centric, business focus

Source: Creaner (2017, September)

11.15   Conclusions

The key factors for being successful in the market are speed and flexibility. Reliability and efficiency are a basic requirement, not a real differentiator in today's world. Without reliability, the quality will suffer, thereby spoiling the customer experience and resulting in loss of market share. Without efficiency, the wastage and consequent high costs will render the firm uncompetitive. High competition, ever-growing demands and expectations of the customers, and constantly evolving technologies, business models, and regulations necessitate an entirely new generation of strategies for winning in the market. Constant innovation is a mantra for profitable growth. Open innovation shows that the organization's architecture needs to be open, to facilitate offering a full and superior range of services and experiences to the customers. This is achieved through a blend of core business and flexibility in collaborating with a variety of partners, such as new market entrants, specialists, and intermediaries.

To accomplish all this, a number of things need to happen. First of all, a change in mindset is in order. A shift from a product-centric to an experience-centric mindset is needed. Design thinking should be used to improve the value of offerings. High automation will not only help minimize costs and bring in superlative efficiency and scale but also provides the much-needed flexibility and the ability to embed intelligence into every level, not just operations. To empower the organization to collaborate and scale seamlessly, platforms are the preferred approach. The entire digital transformation will make data, especially metadata, extremely important. Data analytics, AI, and Machine Learning will become powerful and prominent. All the collaborations will engender the growth of communities and ecosystems. As the complexity of technologies and business models rises, simplification in processes will be preferred. Agile and lean thinking will be the order of the day. A firm can no longer merely deliver what the customer asks for but must constantly study the markets and the competitors and form innovative collaborations to offer outrageously great value and customer experience. For realizing this, the firm needs to constantly build the necessary capabilities and social capital. It's important to build resilience into the strategy and the culture rather than aiming for safety and stability.

An organization's leadership must deliberate upon several trade-offs specific to the organization's business. For example, outsourcing vs. collaboration, open vs. closed innovation, increasing market share vs. increasing margins, and organic vs. inorganic growth.

Evidently, while it is not possible to prescribe one solution or even a one-size-fits-all approach, this book offers many suggestions, guidelines, and frameworks to aid the reader in solving his unique problem. It is hoped that the reader turns to these over and, again to shape and hone strategy, unleash the benefits and become a winner.

> **The Takeaway Box: Your Competition Is Already Hard at Work!**
>
> - Make changes along the directions described in this chapter at the earliest.
> - If you do not initiate the changes today, the organization may rapidly reach a point of no return because the competition will double down and outpace you, adding a larger margin every day.
> - Enter into strategic partnerships to speed up the pace of transformation. This is better than losing out altogether and becoming irrelevant in the market.

References

Afuah, A. (2003). *Innovation Management: Strategies, Implementation and Management* (2nd ed.). Oxford University Press.

Agile Organization. Retrieved from www.agile.org.uk.

Analysys Mason. (2015). *It Is Time for Telcos to Adopt a Platform-Based Business Model*. Retrieved from http://www.analysysmason.com/About-Us/News/Newsletter/telcos-platforms-Jun2015/.

Andriole, S. J. (2018, April). *MIT Sloan Management Review*. Retrieved from https://sloanreview.mit.edu/article/implement-first-ask-questions-later-or-not-at-all/.

Brigwater, A. (2015). What's the Difference Between a Software Product and a Platform. *Forbes*. Retrieved from https://www.forbes.com/sites/adrianbridgwater/2015/03/17/whats-the-difference-between-a-software-product-and-a-platform/#5c00807356a6.

Clarks, J. (2009, June). *The Difference between Platform and Product*. Retrieved from http://jonathanclarks.blogspot.in/2009/06/what-is-difference-between-platform-and.html.

Creaner, M. (2017, September). Telco Digital Transformation: The Conditions, Journeys, and Destinations. *Huawei*. Retrieved from http://www.huawei.com/en/about-huawei/publications/winwin-magazine/plus-intelligence/telco-digital-transformation.

D.School, Stanford's Design Thinking School. Retrieved from https://dschool.stanford.edu/.

Design Management Institute. Retrieved from www.dmi.org.

DVI, Design Value Index. *Design Management Institute*. Retrieved from www.dmi.org.

Hansson, D. H. *The World Needs More Modest, Linear Growth Companies.* Retrieved from https://m.signalvnoise.com/the-world-needs-more-modest-linear-growth-companies-please-make-some-609b5a10a9e0.

Joy, B. *Joy's Law.* Retrieved from https://en.wikipedia.org/wiki/Joy%27s_law_(management).

Knight, F. *Knightian Uncertainty.* Retrieved from https://en.wikipedia.org/wiki/Knightian_uncertainty.

LFN, Linux Foundation for Networking. Retrieved from https://www.linuxfoundation.org/projects/networking/.

MITidm. (2018). *Massachusetts Institute of Technology, Integrated Design and Management.* Retrieved from https://idm.mit.edu.

Moazed, A. (2016, May). *Platform Business Model—Definition.* Retrieved from https://www.applicoinc.com/blog/what-is-a-platform-business-model/.

OCP, Open Compute Project. Retrieved from http://www.opencompute.org/.

Platform Hunt. *The 9 Types of Software Platforms.* Retrieved from https://medium.com/platform-hunt/the-8-types-of-software-platforms-473c74f4536a.

Ries, E. (2011). *The Lean Startup.* Crown Publishing. ISBN 0307887898.

Roberts, E. B. (2002, April). *Innovation: Driving Product, Process, and Market Change* (1st ed.). Jossey-Bass. ISBN-10: 0787962139, ISBN-13: 978-0787962135.

Techopedia. Retrieved from https://www.techopedia.com/definition/3411/platform.

TIP, Telecom Infra Project. (2018). Retrieved from http://telecominfraproject.com/.

Wildcat Ventures. (2017). *The Traction Gap Framework.* Retrieved from http://wildcat.vc/wp-content/uploads/2017/09/Traction-Gap-Framework-9.14.17.pdf.

Zeichick, A. (2018, February 8). *Forbes.* Retrieved from www.forbes.com/sites/oracle/2018/02/05/prediction-80-of-routine-it-operations-will-soon-be-solved-autonomously/#46eec2071b46.

## CHAPTER 12

Fostering Innovation Culture and Exploiting Social Capital: The Cases of T-Hub and RICH

12.1 Introduction

This chapter describes two cases from the real world, T-Hub (Technology Hub) and RICH (Research Innovation Circle of Hyderabad), both based in Hyderabad, India. Both have broken new territory in the landscape.

It is time to update the stereotype of large and successful corporate companies. They may have cutting-edge technology that is research-intensive, but they have made the logic of internal-oriented, centralized R&D somewhat redundant. Outside their firewalls, disruptive startups have progressed in varying degrees, and for everyone to win, the corporate could integrate this innovation into its technology stack. This way, it can reach a higher number of people in as little time as possible, while minimizing cash burn.

In this case, open innovation does not limit itself to two whole units plugging in together. A small group of collaborators could decide to solve a joint problem statement and pool in their talent and resources. A common manifestation of this is adrenaline and caffeine-fueled hackathons.

The idea of "joining forces" in open innovation has a ripple effect across many sectors that reflect in the following ways:

1. In the area of education, it could make all the difference in bringing together the right ideas to the required infrastructure and mentorship.

© The Author(s) 2019
S. Birudavolu, B. Nag, *Business Innovation and ICT Strategies*,
https://doi.org/10.1007/978-981-13-1675-3_12

2. The effect of a tech community's network is amplified in startup incubators and accelerators when academicians, corporates, and the government actively participate in its ongoings. T-Hub, Hyderabad, is a case study that will illustrate this later in the chapter.
3. These companies can interact with one another and beyond transactional terms and explore synergies that were previously not apparent to them.

When ideas become accessible, companies can fail fast, re-innovate, identify mentorship gaps and investment opportunities without losing any time. An efficient process for innovation means we can build the escape velocity to leave this growing period of technology.

The future of innovation lies in the power of collaboration and combination of capabilities to create new economic power.

12.2   Walking into a Real Startup Ecosystem

For this chapter, let us consider T-Hub, Hyderabad, which provides ongoing and relevant instances of open innovation that we can discuss for multiple purposes.

T-Hub's official website states that:

T-Hub is India's largest and fastest startup growth engine catalyzing Innovation, Scale and Deal Flow.

Since November 2015, T-Hub has become India's largest startup campus, catalyzing the growth of startups in a unique public-private partnership with the Government of Telangana, in India.

The T-Hub "Catalyst" at present is an incubator that is one of its kind for growth-stage startups that work in HealthTech, IoT, smart cities, mobility, FinTech, sustainability, and social impact sectors.

Supported by the Department of Science and Technology (DST), T-Hub's Catalyst has a secure network and support from the state government, corporates, investors, and top-class educational institutions. It has earned an enviable position for itself among startups which have regularly been ahead of the innovation curve.

The Catalyst building is home to over 160 growth-stage startups building their next significant solutions to the world. These startups not only get the best infrastructure but also become a part of T-Hub's strong network that actively engages with 1000+ entrepreneurs and community leaders from 14 different countries along with 117 mentors, 24 business support services, among many.

By the end of 2018, in a bid to become the world's second largest startup incubator, T-Hub opens the doors of its second phase, the Reactor building. This centerpiece is 3.5 lakh square feet (the size of 6 football fields) in the middle of HITEC City, the IT cluster in Hyderabad.

Several programs run parallelly at T-Hub, but we shall consider two such programs to examine as use-cases:

- T-Bridge: In the Indian landscape, T-Hub is synonymous for generating and nurturing the next generation of entrepreneurs. T-Bridge was created to invite startups from around the globe to explore the Indian market opportunity with us and allow easy and hassle-free access to India's market and its vast possibilities
- Corporate innovation: T-Hub sees the future of innovation through the lens of entrepreneurs while learning about technology, application trends, and insights to create a culture of innovation to help corporate partners expand opportunities in developing new solutions, products, and services for their businesses faster and more economically

12.3   Case Studies

12.3.1   The India-Israel Innovation Bridge

For startups wading into the deep waters of global commerce, it is not so easy to succeed in the face of uncertainties and risks. Doing business in a new country is difficult due to barriers such as local regulations, infrastructure, mode of doing business, mindsets, language, customs, culture, logistics, geography, and governmental formalities.

Entrepreneurial at heart, T-Hub is the trailblazer for startups looking to expand overseas as well as for startups that are considering India as the next market for their business.

The India-Israel Innovation Bridge is managed by the Israel Innovation Authority and the Department of Industrial Policy and Promotion, Ministry of Commerce and Industry, Government of India. It is a tech platform to facilitate bilateral cooperation between Indian and Israeli startups, tech hubs, corporations, and other key innovation ecosystem players.

The challenge called upon Israeli and Indian startups to come up with solutions to address the critical social issues in the fields of agriculture, water, and digital health. This program had 18 startup winners across all three segments, and they earned the support of both national governments to innovate and expand into each other's markets.

In agriculture, the focus was on reducing post-harvest loss, improving market linkages, enhancing productivity, and increasing farmers' earnings. Yuktix Smart Sensing was one of the startups that lived up to this challenge. Their product, ColdSense, was selected as one of the innovative and affordable solutions for reducing post-harvest losses. As per the Government of India figures, the harvest and post-harvest losses total up to INR 92,651 crores (over 14 billion dollars). The bulk of the post-harvest losses occur due to lack of a good monitoring infrastructure.

The startup Yuktix built a low-cost ICT system for 24/7 monitoring of environmental parameters. Yuktix's computation models for different crops can provide actionable data to cut down waste, improve quality, and boost resource utilization. This helps in boosting income and profitability.

The low-cost (yet hi-tech) sensing systems can impact the lives of millions of people who rely on agriculture for their livelihood. Whereas guesswork currently drives most of the agriculturist's decision-making, battery-powered wireless sensing can be used in the future throughout the agriculture value chain (Singh 2017).

Innovation for water technology revolved around wastewater treatment, desalination, recycling or purifying large water sources, and producing potable drinking water at the point of use. OCEO is a startup founded in 2016 that addresses this problem. It provides an intelligent, sustainable, and low-cost water purification solution to every life across the globe. It has built their first product named OCEO Smart Water Purifier which combines water purification and IoT. This was another successful startup that built a tangible solution over the period of the challenge.

Digital Health invited solutions to health management for NCDs (non-communicable diseases) and low-cost diagnostics in rural areas. Innov4Sight Health and Biomedical Systems is a technology startup that uses ICT, data analytics, biotechnology, biomedical engineering, and bioinformatics and edu-tech in the healthcare sector to reduce errors and improve effective fertility and cancer care. ParSight and Vyabl are its solutions to generate accurate cancer data (Daswani 2017).

12.3.2   Uber Exchange

Ten startups went through a selection process last year and earned the opportunity to have an all-expenses-paid trip to San Francisco by Uber. What they "contested" for among their peers was a chance to co-create with Uber. Uber did choose startups from the transportation sector, but largely remained vertical-agnostic.

Large corporates have a pressing need to accelerate their own innovation and the easiest way to do it is by partnering with disruptive startups. Imagine what would happen when a startup like Gayam Motor Works that manufactures electric bikes and smart autos—the first EV in India with automatic battery swapping system, joins forces with Uber? Uber gets a whole subset of technology and Gayam gets to reach all of Uber's customers. This is the ultimate win-win situation.

There are win-lose situations sometimes, and they are unfortunate occurrences when the corporate giant cannot easily absorb the startup's technology in their stack. The perceived loss here is more towards the corporate.

12.4   The Case of IoT Research Labs

IoTRL is into IoT-powered analytics-driven fleet management, equipping companies with a fleet of vehicles with insights that allow them to increase safety, efficiency, and profitability of fleet operations by 10–15%. This is achieved by using multiple sensors like fuel sensor, temperature sensor, load sensor, RFID/biometric reader and also read the OBD port (on board diagnostics) to monitor critical parameters in real time.

In case of vehicle or operational anomalies, a real-time alert is triggered. The product gives a lot of emphasis to driver behavior which entails harsh acceleration, harsh braking, harsh cornering, overspeeding, excessive vehicle idling which accounts to 20–25% of the vehicles' operational cost in the form of maintenance expenses, operational delays, and so on.

The reports, analytics, and insights enable customers to identify where the inefficiencies lie, with the right information to make data driven decisions.

12.4.1   What Makes IoTRL a Versatile Collaborator

1. The GPS tracking space is very crowded with new players using low-cost hardware and very basic tracking software, which are not very reliable. However, IoTRL only works with fleets that have a fleet size of over 50 vehicles, who realize the value of such a solution.
2. Depending on the industry that we are addressing, and the pain points faced, the right IoT gateways and sensors are deployed.
3. IoTRL provides very industry-specific reports, analytics, and insights as we have a strong team composed of subject matter experts.
4. The accuracy provided by IoTRL's fuel sensors is up to 99% which is unheard of in this industry.
5. The application is very versatile and scalable and can handle lakhs of devices.
6. IoTRL has a command center as a service to improve the effectiveness of the system.
7. Aesthetic design and use of gamification to improve the user experience.
8. The solutions help companies increase efficiency and profitability by 10–15%.

In addition to this, the traction and target market in India itself is huge, with more than 7.1 lakh commercial vehicles added in 2017. Inefficient driving and fuel pilferage lead to a loss of Rs. 70,000 crore each year when considering 7.1 lakh vehicles.

The IoT fleet management market is estimated to be worth 30 billion USD by 2023.

12.5   Open Innovation and Research

The passing of the Bayh-Dole Act (or the Patent and Trademark Law Amendments Act) in the USA in 1980 was an event that changed the route and the rapidity with which research that was conducted in universities impacted the common man. For the first time, universities, non-profits, or individual scientists who were the generators of the Intellectual Property (IP) were allowed to retain ownership of their findings and the economic benefits ensuing thereof. The far-sightedness of this act from the government's perspective shifted focus away from the immediate returns through the sale of the IP in the longer term, and arguably larger, benefits of big picture economic activity—infrastructural investment and job creation. It also set the ball rolling on co-creation in an academic setting by incentivizing the individual innovators to directly engage with industry in the process of commercialization of IP generated using governmental funds directed through universities.

In India, the *Utilization of Public Funded Intellectual Property Bill of 2008*, similar in spirit and intent to the Bayh-Dole Act, was formulated and placed before the Rajya Sabha before being withdrawn due to various shortcomings and a perceived gap between the intent and the likely impact of the bill.

In this context, in the absence of policy and a suitable incentive structure that encourages academic scientists and innovators to directly engage with the industry, there is a case to be made for an independent unbiased entity that serves to reduce the information asymmetry between the four strands of the innovation helix by facilitating the technology transfer process from government-funded R&D labs and institutions.

The Research and Innovation Circle of Hyderabad is an initiative of the Government of Telangana state that was established in early 2017, which became the academic partner to T-Hub.

12.6   Dovetailing with Existing Initiatives

RICH also maintains a complementarity of purpose with T-Hub, another initiative of the Government of Telangana discussed elsewhere in this book. T-Hub mainly caters to startups in the Information Technology sector where design cycles are short, capital investment is low, and the ultimate end product/service is easily scalable. RICH, on the other hand, is active in sectors where the gestation periods are long, capital investment is high, and scaling up from lab scale to the commercial scale is a non-trivial task requiring distinct capabilities. Thus, as a start, RICH works in the following sectors:

1. Food and agriculture
2. Life sciences and pharma
3. Aerospace and defense

The three sectors listed above are ones that have traditionally been a strength of the Hyderabad area, because of which geographic clusters of related industry have already formed over time, with all the necessary capabilities from ideation to scale manufacturing.

RICH has been providing the following business advisory services to its clients in IP sourcing and evaluation for industries, candidate licensee search and evaluation for academia and individual innovators, technology transfer deal structuring, fund raising for client startups and innovators, mentoring and startup scaling advice, and policy advice to the government.

12.6.1   The Year That Was (2017)

As the nature of innovation has changed from many independent companies working on problems and tossing solutions over the wall to downstream stakeholders to a concerted effort towards a common goal from many stakeholders with apparently different aims and incentives, the fundamental nature of businesses that RICH has seen in a year of operation has also changed.

In this collaboration, what started off as three distinct, essentially non-overlapping, verticals meant to reflect the traditional nature of the industries themselves has yielded to a more interdisciplinary picture. Increasingly, innovation appears to be as much, if not more, about the fringes where one business hasn't quite ended and another begins simply because the lack of saturation in these markets means that there is possibly that much more value to be captured.

Given the rate at which the cost of computing power has come down over the last few decades, and also considering the increasing proliferation of smartphones, it is perhaps a foregone conclusion that the common factor underpinning seemingly disparate businesses would be Information Technology (IT) and allied disciplines such as Data Sciences including Artificial Intelligence (AI) and Machine Learning (ML). It is also worth noting at this point that IT and IT-enabled Services (ITeS) seem to play both central and supporting roles depending on application rather than industry, as the traditional models of innovation would have us believe. The other dimension to this revolution of sorts is Communications Technology which has enabled the transfer of data generated by the aforementioned instrumentation systems. The ability to stream data from a remote location has allowed an unprecedented scaling in the number and the distribution of devices from which data can be generated and then processed to produce actionable insights.

For instance, while a lot of research money and effort is being spent on developing newer, more nutritive, and more resilient strains of crops, there are many innovators—large and small, and with less of an agricultural background than a technology background—trying to infuse technology into the field to improve outcomes for the core business of agriculture and its operators.

Satyajeet Mahapatra, originally from the state of Odisha, India, is an engineer settled in Bangalore, India, with more than a decade of IT experience under his belt. Satyajeet founded Exabit Systems a few years ago to build robotic systems to measure weather and soil parameters in a field. Exabit now has a working prototype, one of which is undergoing field trials at the International Crop Research Institute for Semi-Arid Tropics (ICRISAT), a United Nations institute, before large-scale deployment. While Exabit's product is an out-and-out instrumentation system, it has the potential to drastically alter the landscape of microinsurance products which are linked to climatic factors as indicators of crop damage. It could also enable a host of other web or mobile applications which are more aligned towards the core agri-business, but which depend on such weather and soil parameters, such as an app to guide farmers on the next steps to be taken while growing a particular crop—when to sow, when to expect which diseases, when to harvest, and so on.

Similarly, Hyderabad-based Ameya Life, led by Dr. NSD Prasad Rao, is a social enterprise running affordable preventive healthcare programs in schools. While the core business is fundamentally a medical enterprise, the use of Internet of Things (IoT)-based devices to collect data about the students screened and to create a unique cloud-based electronic health record helps to build a comprehensive medical history for each child which in turn improves accuracy of diagnoses as well as treatment outcomes in the event of an illness.

Fourteen Weeks Technologies, another Bangalore-based startup, is making forays into the sustainable power market via its drone-based solar PV health monitoring service. In the midst of a pilot with a large Indian Tier 1 solar PV company, the usage of remotely controlled drones coupled with data science tools to quickly, correctly and consistently locate, identify, and assess faults in solar PV panels allows the company to save close to 80% on time and 60% on manpower cost when compared with an equivalent manual approach.

12.6.2   The Way Forward

The strategy ahead for RICH needs to be dealt with on two levels—at the strategic level and at the operational level.

At the strategic level, through various engagements, RICH has helped cross-link the various strands of the quadruple helix—the government, the industry, academia, and individual innovators. The nature of the interactions between each of these stakeholders can be seen to be taking various forms depending on the task at hand, such as:

1. Governments giving innovators access to patient data (while abiding by privacy regulations) to improve outcomes for diagnostic and prescriptive protocols
2. Universities and research labs giving individual innovators access to specialized equipment in order to drive utilization for the labs themselves and drive down capital costs for the innovators
3. Large corporates giving individual innovators access to their networks to help them scale up faster while also building a strong vendor base for themselves

RICH has been enabling and strengthening these cross-links across various industries while also looking for newer models of co-creation for the future.

The next piece of the puzzle that remains to be solved, however, is how this can be turned into a playbook or a replicable model that can quickly be made operational in different settings, perhaps in a different geography or catering to very niche industries, and so on. There is also the related question of how to measure the impact of such an initiative, especially when some of the businesses supported are social enterprises and traditional financial metrics may present an incomplete picture at best.

At the operational level, fortunately, the way ahead is unequivocally clear. The tools used in all the previous examples are inherently discipline-agnostic which is also what endows them with the power to scale diverse businesses so uniformly across the board. What is immediately needed, and what is already underway, is to focus on these fundamental building blocks of systems rather than just systems as a whole and to build capability in these building block technologies such as IoT, AI, and so on since the quicker wins for organizations like RICH serving traditional and saturated businesses are likely to be at the junction of the old and new, where impact or transformation is also likely to be that much more dramatic.

12.7 FROM CLOSED TO OPEN INNOVATION

Is innovation dead by opening it to a wider audience? Hardly, as indicated by the latest advances in life sciences, including breakthroughs in medicine and healthcare. Is internal R&D no longer as relevant? The answer lies in an underlying shift in how companies spawn new ideas and bring them to market. In the old model of closed innovation, firms adhered to the mindset that successful innovation requires control (Chesbrough 2001). In other words, companies must do their own research to generate and develop their own ideas. They would then take full ownership and manufacture, market, distribute, and service, all on their own. This traditional approach calls for self-reliance: If you want something done right, you've got to do it yourself.

Partnerships are essentially a collaboration between companies. Ventures can be described as investments in early stage startups, as in the case of the *Uber Exchange* program. These investments can bring revenue in case of exits but also provide access to new technologies that Uber can learn and benefit from.

Uber offers these startups help in the form of an initial investment and some resources. The idea is that the products built by the internal startups could be included in Uber's product portfolio, or just serve as a learning experience for both Uber and the startup.

While the world saw an unprecedented change due to disruptive innovations, India also saw an impactful change in the startup ecosystem. From having a handful of VCs and almost no incubators in 2006, today we have over 50 venture capital investors, 5000 angel investors (small and big), and 20+ high-quality incubators, both private and public. Also, India saw only 12 lakh companies getting incorporated in the last 70 years, while today, we add 1 lakh every year, that is, 10 new companies are started in India every hour! In an era like this, the mentality to co-create keeps the hazards of the "survival of the fittest" instinct at bay and enables the doing of good work.

REFERENCES

Chesbrough, H. (2001). *Open Innovation*, Open Innovation Community. Retrieved from http://openinnovation.net/about-2/open-innovation-definition/.

Daswani, S. (2017, December). *Meet the Winners of The India-Israel Innovation Challenge*, Inc42. Retrieved from https://inc42.com/buzz/india-israel-innovation-challenge-winners/.

Singh, S. (2017, December 28). *Startup India India-Israel Innovation Challenge Results Are Out*. Yuktix Technologies Blog. Retrieved from http://blog.yuktix.com/.

GLOSSARY

AGR Annual gross revenue

AI Artificial Intelligence

API Application programming interface

Application When used as a noun, it refers to software applications

AR Augmented Reality

ARPU Average revenue per user, of a Telecom operator

B2B Business to business

B2C Business to consumer

B2B2C Business to business to consumer

BIS Bureau of Indian Standards

B/OSS Combined name for BSS and OSS

BSS Business support systems

BTC Bitcoin

CME Communications, media, and entertainment industry; also known as ICT, IMT, TMT, and Telecom

CDMA Code-division multiple access

Coopetition or Co-opetition Collaboration between business competitors, in the hope of mutually beneficial results

CPA Content provider access

CPE Customer premises equipment; this includes servers, routers, session border controllers, and so on

© The Author(s) 2019   351
S. Birudavolu, B. Nag, *Business Innovation and ICT Strategies*,
https://doi.org/10.1007/978-981-13-1675-3

CSP Communication service provider; also known as Telco, Telecom provider, Telecom operator. Alternately, depending on the context, it could also mean cloud service provider

DCB Direct carrier billing

DoT Department of Telecom in India

DVI Design Value Index

EBITDA Earnings before interest, taxes, depreciation, and amortization

EU European Union

EUGDPR European Union's General Data Protection Regulation

Firm Telco firm

Forum Unless otherwise understood in a different context, in this thesis it refers to open innovation-based ICT alliances forum or consortium

GATT General Agreement on Tariffs and Trade

GSM Global System for Mobile Communications

HHI Herfindahl-Hirschman Index, a standard measure of competitive intensity in a country. This is sector specific

ICT Information Communications Technology industry, which is telecommunications' extended ecosystem and industry. Also known as CME, or IMT, TMT, or Telecom

IMT Information Media Technology; also known as ICT, TMT, CME, Telecom

Industry Refers to the ICT industry, unless otherwise specified

InfoSec Information Security

Intrapreneur A person within a large corporation who takes direct responsibility for turning an idea into a profitable finished product through assertive risk-taking and innovation. This definition has been taken from The American Heritage Dictionary of the English Language, and the term is widely used in the industry today. The intrapreneur has the enterprise and thinking of an entrepreneur but is employed in an organization and applies his talents, skill, and effort for the organization

Intrapreneurship or Intrapreneurial Attitude Embodying the spirit and thinking of an intrapreneur. Actively playing the role of an intrapreneur

IoT Internet of Things

IP Internet Protocol

ITA Information Technology Agreement

KMO Kaiser-Meyer-Olkin statistical measure of sampling adequacy used in principal components analysis

M2M Machine-to-machine

MDS Mobile data services

MeitY Ministry of Electronics and Information Technology in India

ML Machine Learning

MPLS Multiprotocol Label Switching

MR Mixed reality

MVAS Mobile value-added services

MVNO Mobile virtual network operator

NASSCOM The National Association of Software and Services Companies is a non-profit trade association of Indian Information Technology and business process outsourcing industry

NGOSS Next-generation OSS, from TMF. This has been superseded by TMF's Frameworx

NFV Network function virtualization

NTP New Telecom Policy in India

OI Open innovation

OSS Operations support systems

PCA Principal components analysis in statistics

PPP Purchasing power parity

PSTN Public switched telephone network

Regression Linear regression in statistics

Revenue-Split or Revenue-Sharing Refers to the percentage of revenue that the Telco shares (i.e. gives away) to the VAS/MDS partner out of the total revenue yield from a service launched jointly with the VAS/MDS partner

SDN Software-defined networking

SIP Session Initiation Protocol

Telco Telecom operator; otherwise known as Telecom service provider or CSP. Examples of Telcos are Vodafone, AT&T, Bharti Airtel, Telefonica, Verizon, China Mobile, Softbank, KDDI, Deutsche Telekom, Sprint, Singtel, and so on

Telecom Telecommunications industry; also known as ICT, TMT, CME, IMT

TMF TeleManagement Forum

TMT Telecommunications, media, technology industry; also known as ICT, IMT, CME, Telco, and telecommunications

TOGAF The Open Group Architecture Framework, from the Open Group

TRAI The Telecom Regulatory Authority of India

VAS Value-added services, otherwise called MVAS in this thesis

VIF Variance inflation factor; in the results of linear regression, VIF is inspected for the individual regressors to check the presence of multi-collinearity. A VIF of less than 2 indicates no multi-collinearity

VoIP Voice over IP

VR Virtual Reality

WEF World Economic Forum

WTO World Trade Organization

XaaS An acronym for "X as a Service", where "X" is a generic placeholder for software, process, knowledge, network, and so on, and the acronym expands to "software as a service", "process as a service", and so on. These are usually cloud-based service offerings

XR Extended Reality

INDEX

© The Author(s) 2019
S. Birudavolu, B. Nag, *Business Innovation and ICT Strategies*,
https://doi.org/10.1007/978-981-13-1675-3